Fourier Analysis

Fourier Analysis

Roger Ceschi
Jean-Luc Gautier

iSTE

WILEY

First published 2017 in Great Britain and the United States by ISTE Ltd and John Wiley & Sons, Inc.

ISTE Ltd
27-37 St George's Road
London SW19 4EU
UK

www.iste.co.uk

John Wiley & Sons, Inc.
111 River Street
Hoboken, NJ 07030
USA

www.wiley.com

Library of Congress Control Number: 2016957604

British Library Cataloguing-in-Publication Data
A CIP record for this book is available from the British Library
ISBN 978-1-78630-109-3

Contents

Preface

This book is presented as a collection of exercises and their solutions that aim to help undergraduate and postgraduate students studying signal processing.

The exercises are originally from a book by B.P. Lathi entitled *Signals, Systems and Communications*, published by John Wiley.

In this book, we concern ourselves only with questions surrounding deterministic signals.

This book is not intended to be a lesson and thus we deemed it unnecessary to present all of the mathematical steps for demonstrating the solutions. We also intentionally ignored many of the inherent difficulties that stem from the formalism of the distributions we encounter.

We hope our mathematician readers understand some of the ambiguities or even uncertainties that may inevitably occur.

We highly encourage any readers interested in additional information and details surrounding the demonstrations to refer to the books included in the Bibliography.

Roger CESCHI
Jean-Luc GAUTIER
November 2016

1

Fourier Series

1.1. Theoretical background

1.1.1. *Orthogonal functions*

1.1.1.1. *Orthogonal vectors*

Let us consider a vector space with n dimensions and vectors $\mathbf{x_1}, \mathbf{x_2}, \ldots \mathbf{x_n}$ as the orthogonal basis.

$$\mathbf{x_i^\dagger} . \mathbf{x_j} = \begin{cases} 0 & \text{if } i \neq j \\ k_i & \text{if } i = j \end{cases} \qquad [1.1]$$

k_i is the norm for vector $\mathbf{x_i}$.

$\mathbf{x_i^\dagger} = \mathbf{x}^{\mathbf{T}*}$ is the Hermitian vector (conjugate and transposed) of vector $\mathbf{x_i}$.

This collection of vectors is assumed to be complete once there is no way of finding any more values for $\mathbf{x_k}$ such as $\mathbf{x_k^\dagger} . \mathbf{x_i} = \mathbf{0}$

Consider vector $\mathbf{A}$ within this space and $A_1, A_2, \ldots A_n$ this vector's components in relation to n basis vectors. If the basis is complete, we can pose:

$$\mathbf{A} = \sum_{i=1}^{i=n} A_i \mathbf{x_i} \qquad [1.2]$$

We obtain coefficients A_i using to the following relation:

$$A_i = \frac{\mathbf{A}^\dagger.\mathbf{x_i}}{\mathbf{x_i^\dagger}.\mathbf{x_i}} = \frac{\mathbf{A}^\dagger.\mathbf{x_i}}{k_i} \quad [1.3]$$

1.1.1.2. *Vector-function analogy*

Consider two functions $f_i(t)$ and $f_j(t)$ that are orthogonal on interval $[t_1, t_2]$ if:

$$\int_{t_1}^{t_2} f_i(t) f_j^*(t) dt = \begin{cases} o & \text{if } i \neq j \\ k_i & \text{if } i = j \end{cases} \quad [1.4]$$

k_i is the squared norm of function $f_i(t)$.

The vector space is complete if we can no longer find any further functions $f(t)$ that are orthogonal to the previous ones.

When this space is complete and infinite, we can establish an exact representation of any function as a series on interval $[t_1, t_2]$:

$$g(t) = \sum_{i=1}^{i=\infty} C_i f_i(t) \quad t \in [t_1, t_2] \quad [1.5]$$

The coefficients of the decomposition are found using the following relation:

$$C_i = \frac{\int_{t_1}^{t_2} g(t) f_i^*(t)}{\int_{t_1}^{t_2} f_i(t) f_i^*(t)} = \frac{\int_{t_1}^{t_2} g(t) f_i^*(t)}{k_i} \quad [1.6]$$

1.1.2. *Fourier Series*

1.1.2.1. *Trigonometric series*

Functions $\{\cos n\omega_0 t\}$ and $\{\sin n\omega_0 t\}$ form, on interval $[t_0, t_0+T]$, an infinite complete collection of orthogonal functions with $T = \dfrac{2\pi}{\omega_0}$.

It is thus possible to represent a function $f(t)$ on interval $[t_0, t_0+T]$:

$$f(t) = a_0 + \sum_{n=1}^{n=\infty} a_n \cos n\omega_0 t + \sum_{n=1}^{n=\infty} b_n \sin n\omega_0 t \quad t \in [t_0, t_0+T] \qquad [1.7]$$

a_0 is the mean value of $f(t)$ on interval $[t_0, t_0+T]$

The coefficients are found using the following relations:

$$\begin{cases} a_0 = \dfrac{1}{T}\displaystyle\int_{t_0}^{t_0+T} f(t)\,dt \\ a_n = \dfrac{2}{T}\displaystyle\int_{t_0}^{t_0+T} f(t)\cos n\omega_0 t\;dt \\ b_n = \dfrac{2}{T}\displaystyle\int_{t_0}^{t_0+T} f(t)\sin n\omega_0 t\;dt \end{cases} \qquad [1.8]$$

1.1.2.2. *Exponential series*

Functions $\{e^{jn\omega_0 t}\}$ form an infinite complete collection of orthogonal functions on interval $[t_0, t_0+T]$.

On this same interval, a function $f(t)$ would be noted:

$$f(t) = \sum_{n=-\infty}^{n=\infty} C_n e^{jn\omega_0 t} \quad t \in [t_0, t_0+T] \qquad [1.9]$$

The coefficients for the decomposition are found using the following:

$$C_n = \frac{1}{T}\int_{t_0}^{t_0+T} f(t)e^{-jn\omega_0 t}\,dt \qquad [1.10]$$

In the case where $f(t)$ is real, we know that the following is true:

$$C_n = C_{-n}^* \qquad [1.11]$$

Thus $|C_n| = |C_{-n}|$ et $Arg(C_n) = -Arg(C_{-n})$

This gives us an example of Hermitian symmetry.

1.1.2.3. *Relations between two different forms of series*

The following relations allow us to change between one form of Fourier series and another:

$$\begin{cases} a_0 = C_0 & \\ a_n = c_n + C_{-n} & b_n = j(C_n - C_{-n}) \\ C_n = \frac{1}{2}(a_n - jb_n) & C_{-n} = \frac{1}{2}(a_n + jb_n) \end{cases} \qquad [1.12]$$

There is an alternative form for the trigonometric series:

$$\begin{cases} f(t) = a_0 + \sum_{n=1}^{n=\infty} A_n \cos(n\omega_0 t + \phi_n) \\ A_n = \sqrt{a_n^2 + b_n^2} \ \text{ et } \ tg\phi_n = -\frac{b_n}{a_n} \end{cases} \qquad [1.13]$$

Table 1.1 presents a recap of these different formulas.

<table>
<tr><th>Series form</th><th>Calculation of coefficients</th></tr>
<tr><td>$f(t)=\sum\limits_{n=-\infty}^{n=\infty} C_n e^{jn\omega_0 t}$</td><td>$C_n=\frac{1}{T}\int\limits_{t_0}^{t_0+T} f(t)e^{-jn\omega_0 t}\,dt$</td></tr>
<tr><td>$f(t)=a_0+\sum\limits_{n=1}^{n=\infty} a_n\cos n\omega_0 t+\sum\limits_{n=1}^{n=\infty} b_n \sin n\omega_0 t$
$=a_0+\sum\limits_{n=1}^{n=\infty} A_n\cos\left(n\omega_0 t+\phi_n\right)$</td><td>$a_0=\frac{1}{T}\int\limits_{t_0}^{t_0+T} f(t)\,dt$
$a_n=\frac{2}{T}\int\limits_{t_0}^{t_0+T} f(t)\cos n\omega_0 t\,dt$
$b_n=\frac{2}{T}\int\limits_{t_0}^{t_0+T} f(t)\sin n\omega_0 t\,dt$</td></tr>
<tr><th colspan="2">Formulas for switching between forms</th></tr>
<tr><td>$a_0=C_0$</td><td></td></tr>
<tr><td>$a_n=C_n+C_{-n}$</td><td>$b_n=j\left(C_n-C_{-n}\right)$</td></tr>
<tr><td>$C_n=\frac{1}{2}\left(a_n-jb_n\right)$</td><td>$C_{-n}=\frac{1}{2}\left(a_n+jb_n\right)$</td></tr>
<tr><td>$A_n=\sqrt{a_n^2+b_n^2}=2\left|C_n\right|$</td><td>$\phi_n=Arg\left(C_n\right)=-Arctg\left(\frac{b_n}{a_n}\right)$</td></tr>
</table>

Table 1.1. *Recap of Fourier series formulas*

1.1.3. *Periodic functions*

A periodic function with a period T is a function that repeats itself identically every T seconds.

$$f(t+nT)=f(t)\quad \forall t \qquad n\in Z \tag{1.14}$$

Consider a periodic function of period T represented on interval $\left[t_0,t_0+T\right]$ by a Fourier series, the representation remains valid regardless of whether $t\in\left]-\infty,\infty\right[$.

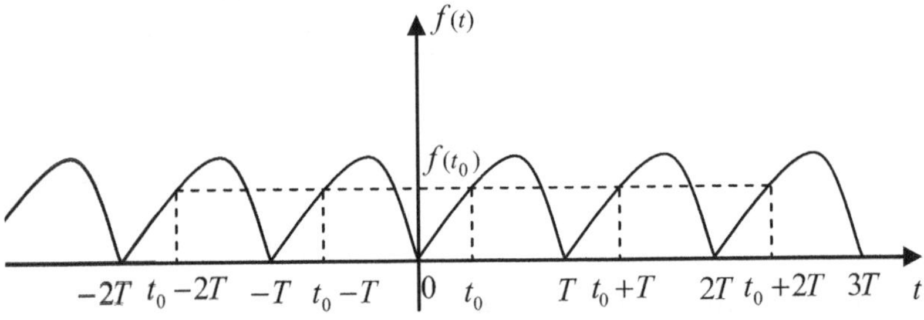

Figure 1.1. *Periodic function*

The value for t_0 is thus irrelevant. In practice, we often set $t_0 = 0$ or $t_0 = -\frac{T}{2}$, the integrals of formulas 8 and 10 thus become:

$$\int_0^T \text{ and } \int_{-T/2}^{T/2}$$

1.1.4. *Properties of Fourier series*

1.1.4.1. *Time domain translation*

If $f(t)$ is represented by the following Fourier series:

$$f(t) = \sum_{n=-\infty}^{n=\infty} C_n e^{jn\omega_0 t}$$

When translated in time, function $f(t-\tau)$ can be noted as follows:

$$\begin{cases} f(t-\tau) = \sum_{n=-\infty}^{n=\infty} C_n' e^{jn\omega_0 t} \\ C_n' = C_n e^{-jn\omega_0 \tau} \end{cases} \qquad [1.15]$$

1.1.4.2. *Even functions*

If $f(t)$ is an even function, that is $f(t)=f(-t)$, its decomposition into a Fourier series will not contain any sine values.

$$f(t)=a_0+\sum_{n=1}^{n=\infty} a_n \cos n\omega_0 t$$

It is also possible to cut the integration interval by half:

$$\begin{cases} a_0 = \dfrac{2}{T}\displaystyle\int_0^{T/2} f(t)\,dt \\ a_n = \dfrac{4}{T}\displaystyle\int_0^{T/2} f(t)\cos n\omega_0 t\; dt \end{cases} \qquad [1.16]$$

1.1.4.3. *Odd functions*

If $f(t)$ is an odd function, i.e. $f(t)=-f(-t)$, its decomposition into a Fourier series will only contain sine values.

$$\begin{cases} f(t) = \displaystyle\sum_{n=1}^{n=\infty} b_n \sin n\omega_0 t \\ b_n = \dfrac{4}{T}\displaystyle\int_0^{T/2} f(t)\sin n\omega_0 t\; dt \end{cases} \qquad [1.17]$$

An absence of function parity can be masked by the mean value of function $f(t)$. Therefore, we must analyze this property in the case of function $f(t)-a_0$.

1.1.4.4. *Rotational symmetry*

In this case, function $f(t)$ is composed of two identical half-periods of opposing signs: $f(t\pm\frac{T}{2})=-f(t)$.

The decomposition into a Fourier series contains only odd harmonics: n odd.

This property can also be masked by the mean value of the function.

1.1.5. *Discrete spectra. Power distribution*

A periodic function $f(t)$ has a frequency spectrum which provides it with a representation on the frequency domain. This spectrum only exists for discrete values of ω. It is either a discrete or a line spectrum.

The amplitude of each spectral line is proportional to the value set by the function.

1.1.5.1. *Physical spectrum. Complex spectrum*

The physical spectrum represents values A_n and ϕ_n. It corresponds to positive values of frequency.

The complex spectrum represents C_n. There are therefore two spectra, one is the amplitude spectrum and the other is the phase spectrum. This highlights negative pulses; the amplitude of the spectral lines at $n\omega_0$ is actually the combination of all the spectral lines at $\pm n\omega_0$.

In a case where function $f(t)$ is real, $C_n = C^*_{-n}$ the amplitude spectrum will be symmetric, while the phase spectrum will be antisymmetric to the origin of the frequency.

The phase spectra are identical while the amplitude spectra are linked through the two following equations $\begin{cases} C_0 = a_0 \\ 2|C_n| = A_n \end{cases}$

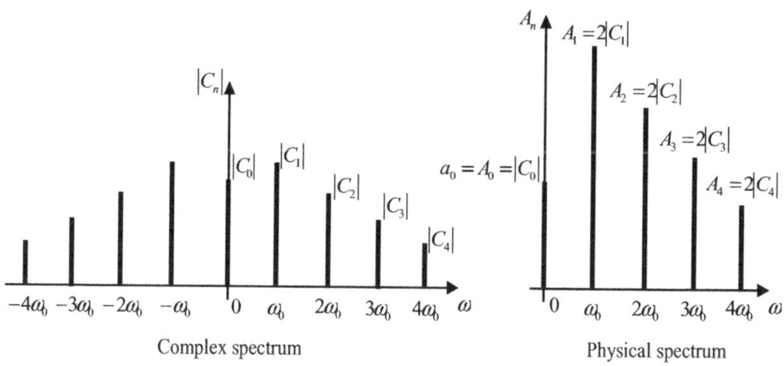

Figure 1.2. *Physical spectrum and complex spectrum*

1.1.5.2. *Power spectrum*

The mean quadratic value of a periodic function $f(t)$ or root mean square (RMS), is given by the following equation:

$$F'^2_{eff} = \frac{1}{T}\int_0^T f^2(t)\, dt$$

This value is directly linked to the power of the signal:

$$F^2_{eff} = \frac{1}{T}\int_0^T f^2(t)dt = \sum_{n=-\infty}^{n=+\infty} |C_n|^2 \qquad [1.18]$$

Parseval's theorem demonstrates how the power contained within a signal is the sum of the powers contained in all the spectral lines of its Fourier decomposition.

It is possible to trace a power spectrum representing $|C_n|^2$.

1.2. Exercises

The exercises presented in this chapter come from B.P. Lathi [LAT 66].

1.2.1. *Exercise 1.1. Examples of decomposition calculations*

For each of the following periodic functions, calculate a Fourier decomposition and trace the outline of the complex spectrum.

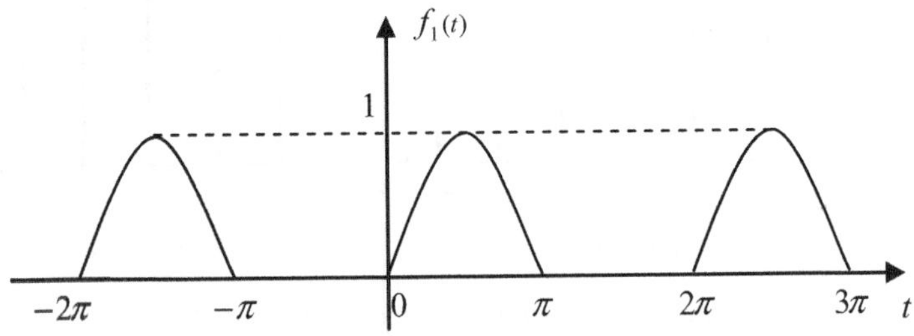

Figure 1.3. *Sinusoidal arc function*

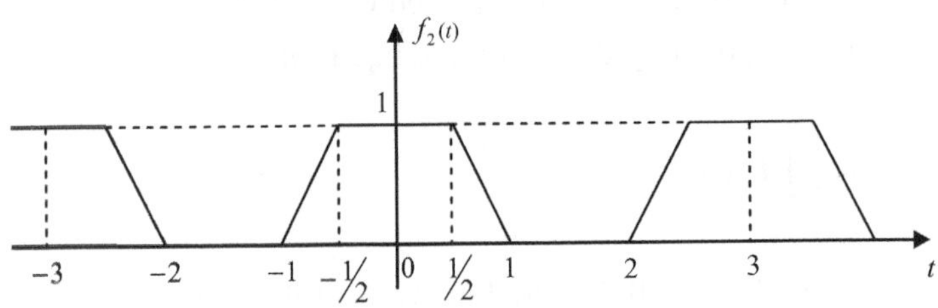

Figure 1.4. *Trapezoid function*

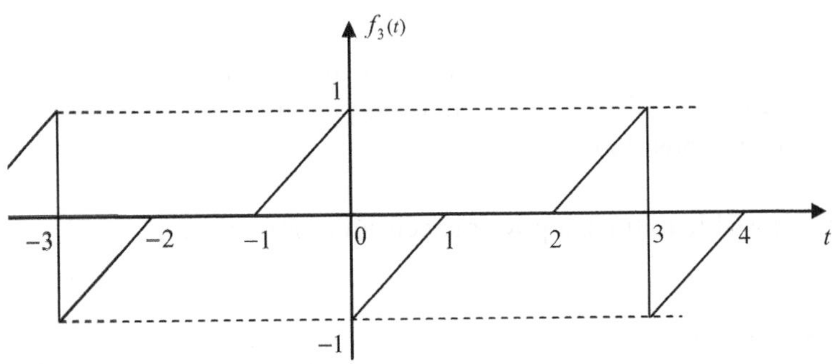

Figure 1.5. *Saw tooth function*

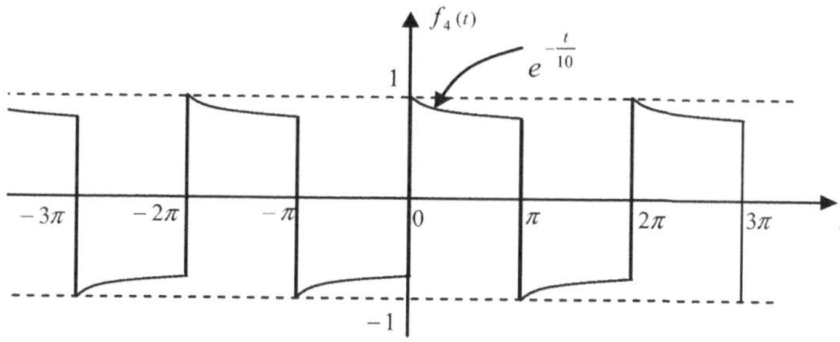

Figure 1.6. *Exponential function.*

In the case of Figure 1.6, what limit does this decomposition approach when the exponential time constant becomes infinite?

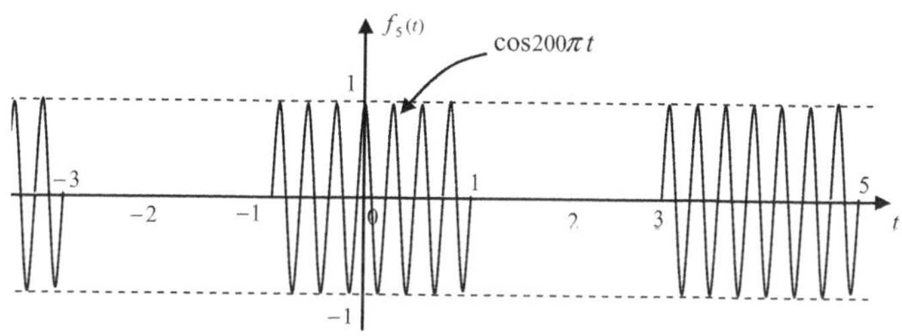

Figure 1.7. *Modulated Sinusoidal function*

1.2.2. *Exercise 1.2*

Consider the following rectangular function:

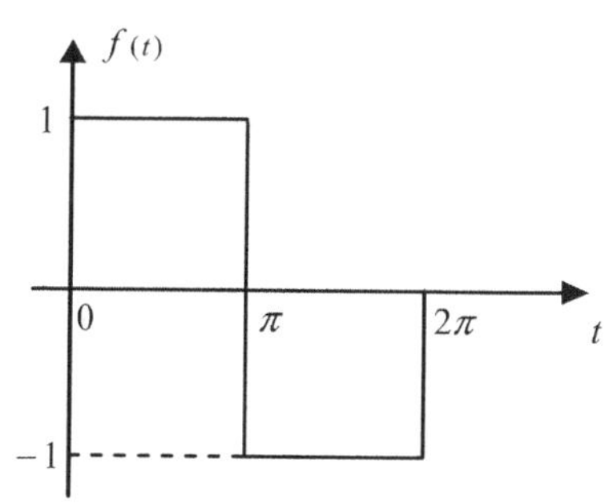

– Demonstrate that on interval $[0, 2\pi]$, $f(t)$ is orthogonal to $\cos nt$ with n integer number. What conclusion would you draw from this regarding the Fourier decomposition of $f(t)$?

– Demonstrate that on the same interval, error function $f_e(t) = f(t) - \frac{4}{\pi}\sin t$ is orthogonal to $\sin t$. Explain this result.

1.2.3. *Exercise 1.3*

Consider two functions $f_1(t)$ and $f_2(t)$ that are orthogonal on interval $[t_1, t_2]$.

We define the energy of a function on this interval with the following equation:

$$E = \int_{t_1}^{t_2} f^2(t)dt$$

Demonstrate that the energy contained in the sum, or difference, of these two functions $f_1(t)$ and $f_2(t)$ is the sum of the energies contained in each of these functions.

1.2.4. *Exercise 1.4*

Demonstrate that for an even periodic function, the coefficients of the Fourier decomposition are real and pure imaginary for an odd function.

1.2.5. *Exercise 1.5*

– Give an example of two periodic functions that have the same Fourier series with the exception of component a_0.

– What are the differences between Fourier developments of functions that are identical in everything other than their periods? Make a comparison between a function that has a period of 2π

seconds with another function that is identical except that it has a period of π milliseconds.

1.2.6. *Exercise 1.6. Decomposing rectangular functions*

– Calculate the Fourier decomposition of the following function:

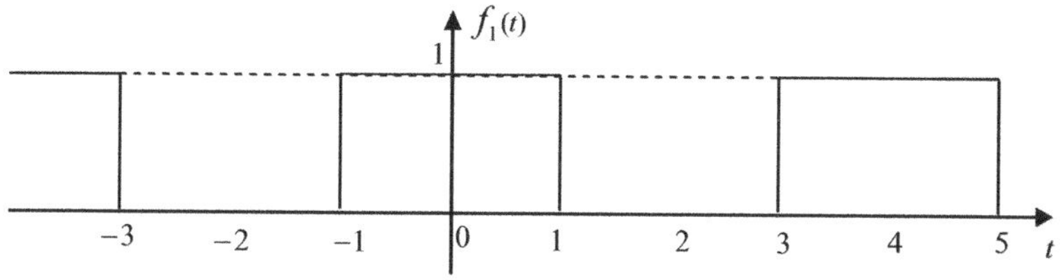

– Using this result, establish in the simplest way possible the decompositions of the following periodic functions:

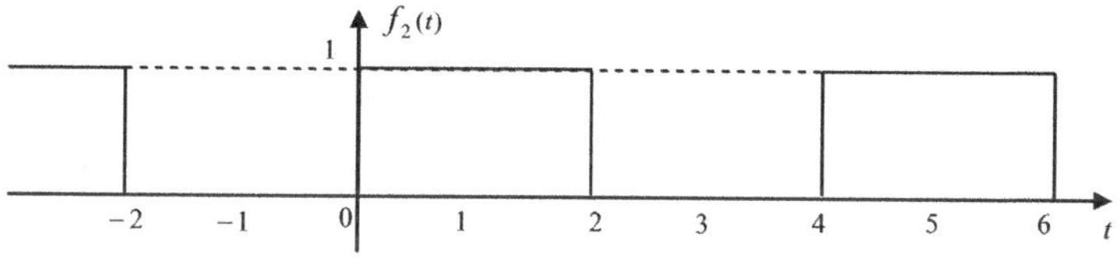

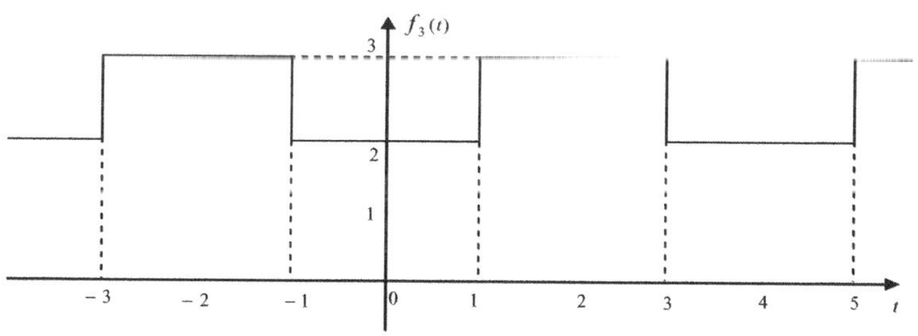

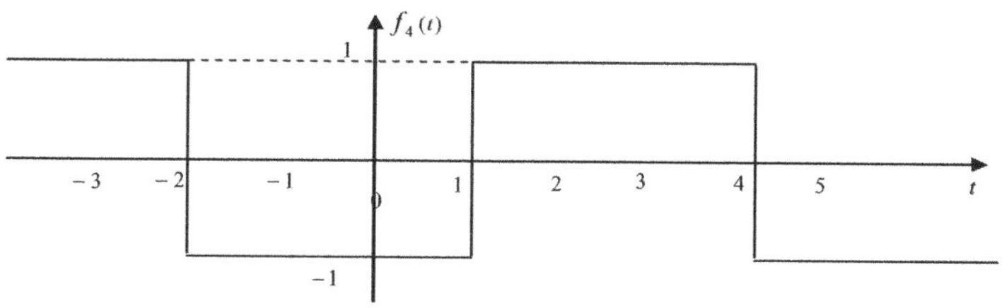

1.2.7. *Exercise 1.7. Translation and composition of functions*

Consider the following function $f(t)$:

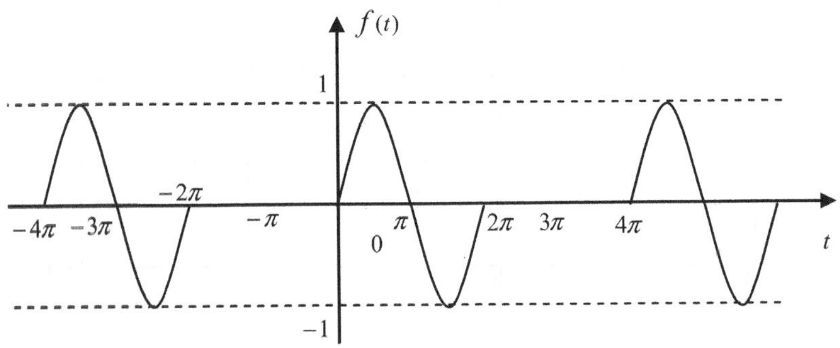

– directly calculate the Fourier decomposition of $f(t)$;

– perform a translation of π to the left, or $f_i(t) = f(t+\pi)$. Decompose $f_i(t)$ into a Fourier series and then find the result for $f(t)$ starting from $f_i(t)$;

– same exercise but this time performing a translation of π to the right, $f_i^{'}(t) = f(t-\pi)$;

– consider the following function $f_1(t)$:

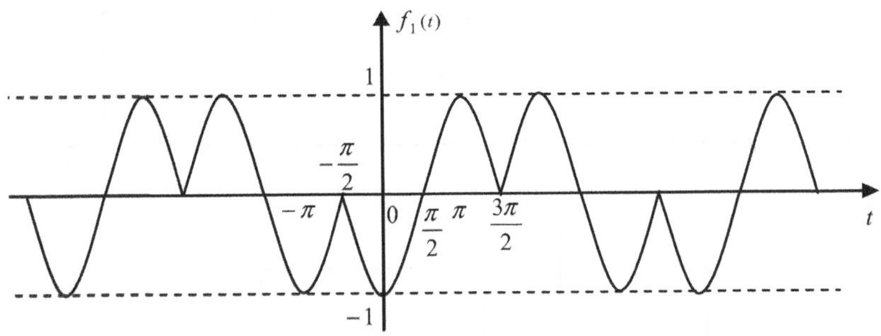

Demonstrate that $f_1(t)$ can be expressed as a function of $f(t)$. Use the previous results to determine the decomposition of $f_1(t)$.

1.2.8. *Exercise 1.8. Time derivation of a function*

Demonstrate that if function $f(t)$ is periodic and derivable, function $f'(t) = \dfrac{df(t)}{dt}$ is also periodic.

From this, show that if $f(t) = \sum_{n=-\infty}^{n=\infty} C_n e^{jn\omega_0 t}$, the Fourier decomposition of the derived function is:

$$\frac{df(t)}{dt} = \sum_{n=-\infty}^{n=\infty} jn\omega_0 C_n e^{jn\omega_0 t}$$

1.2.9. *Exercise 1.9. Time integration of functions*

Consider a periodic function $f(t)$. Demonstrate that integral function $F(t) = \int f(t)dt$ is periodic on condition that the mean value of $f(t)$ be 0.

In this case, demonstrate that $F(t) = \sum_{n=-\infty}^{n=\infty} \dfrac{C_n}{jn\omega_0} e^{jn\omega_0 t}$.

1.2.10. *Exercise 1.10*

Consider the following periodic function $f(t)$:

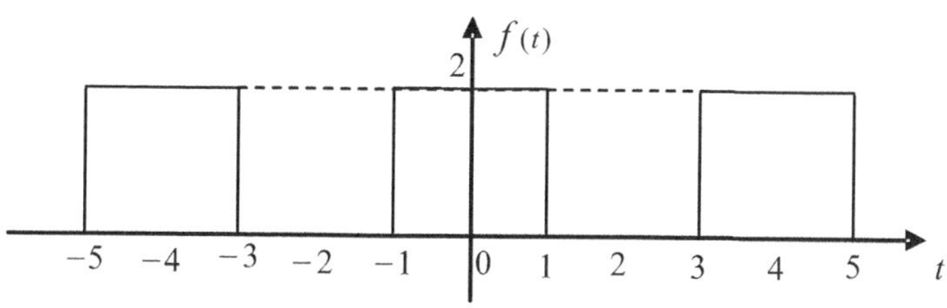

With the following Fourier decomposition:

$$f(t) = \sum_{n=-\infty}^{n=+\infty} \frac{\sin n\frac{\pi}{2}}{n\frac{\pi}{2}} e^{jn\frac{\pi}{2}t}$$

Using the previous function, establish the decomposition of the following function $g(t)$:

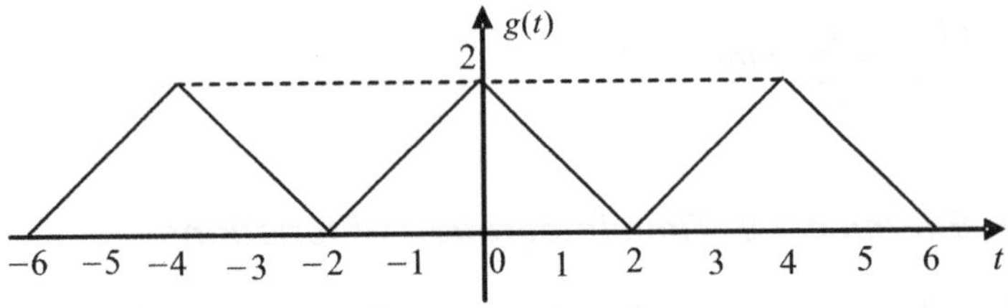

We can use the results from exercise 1.9.

1.2.11. *Exercise 1.11. Applications in electronic circuits*

Let us consider a periodic voltage $v(t)$ of period $T = 1s$, applied to the following circuit:

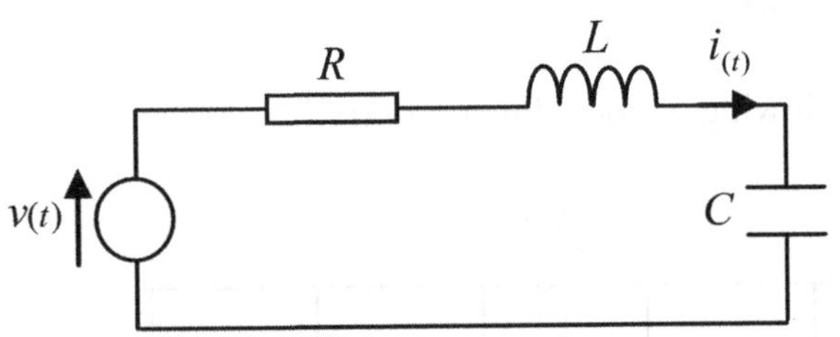

$R = 0.01\Omega$

$L = 1H$

$C = \frac{1}{16\pi^2} F$

Find the Fourier decomposition of current $i(t)$ when $v(t)$ takes the following forms. What can you say about these results?

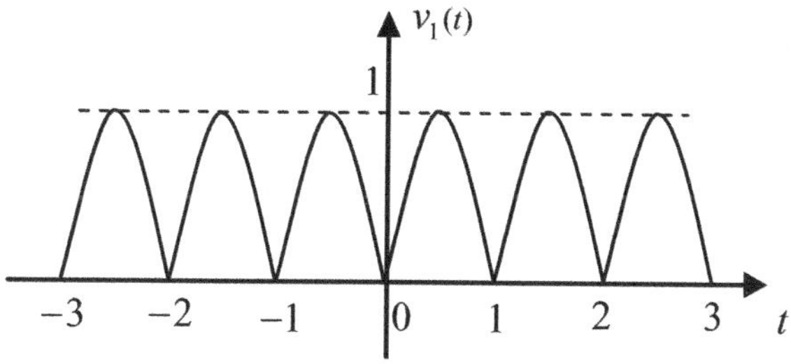

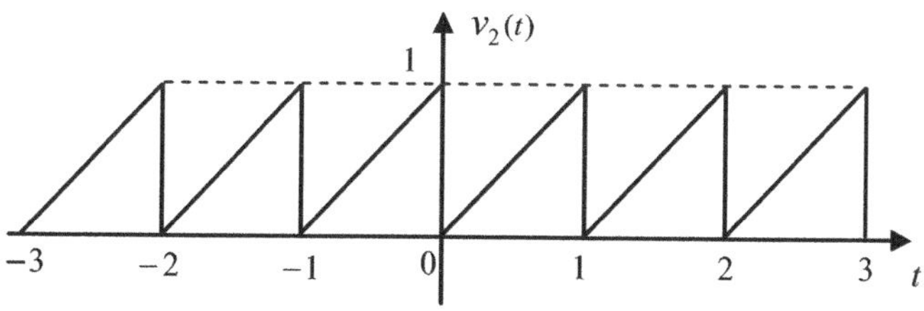

1.3. Solutions to the exercises

1.3.1. *Exercise 1.1. Examples of decomposition calculations*

– Function $f_1(t)$:

The analytical form of function $f_1(t)$ of Figure 1.3 during a single period is:

$$f_1(t) = \begin{cases} \sin t & t \in [0, \pi] \\ 0 & t \in [\pi, 2\pi] \end{cases}$$

The period here is $T = 2\pi$, the fundamental pulse is $\omega_0 = 1$.

Let us find the decomposition in trigonometric form:

$$a_0 = \frac{1}{T}\int_0^T f_1(t)dt = \frac{1}{2\pi}\int_0^\pi \sin t dt$$

or: $$a_0 = \frac{1}{\pi}$$

$$a_n = \frac{2}{T}\int_0^T f_1(t)\cos n\omega_0 dt = \frac{2}{2\pi}\int_0^\pi \sin t \cos nt dt$$
$$= \frac{1}{2\pi}\int_0^\pi [\sin(n+1)t - \sin(n-1)t]dt$$

Hence $a_n = \frac{-1}{2\pi}\left[\frac{\cos(n+1)t}{n+1} - \frac{\cos(n-1)t}{n-1}\right]_0^\pi$ for $n \neq 1$

or: $$a_n = \frac{1}{\pi}\frac{1+\cos n\pi}{1-n^2}$$

If n is even, we can set $n = 2p$, thus: $a_{2p} = \frac{1}{\pi}\frac{2}{1-4p^2}$

If n is odd, then $n = 2p+1$, thus: $a_{2p+1} = 0$

We now perform this calculation again with $n = 1$:

$$a_1 = \frac{1}{\pi}\int_0^\pi \sin t \cos t dt = \frac{1}{2\pi}\int_0^\pi \sin 2t dt = \frac{-1}{\pi}\left[\frac{\cos 2t}{2}\right]_0^\pi = 0$$

We therefore deduce that: $a_{2p+1} = 0 \quad \forall p$

$$b_n = \frac{2}{T}\int_0^T f_1(t)\sin n\omega_0 dt = \frac{2}{2\pi}\int_0^\pi \sin t \sin nt dt$$
$$= \frac{1}{2\pi}\int_0^\pi [\cos(n+1)t - \cos(n-1)t]dt$$
$$= \frac{1}{2\pi}\left[\frac{\sin(n+1)t}{n+1} - \frac{\sin(n-1)t}{n-1}\right]_0^\pi \quad n \neq 1$$

Therefore: $\boxed{b_n = 0 \quad \forall n \neq 1}$

Using the calculation for b_1:

$$b_1 = \frac{1}{\pi}\int_0^{\pi} \sin^2 t dt \text{ or } b_1 = \frac{1}{2}$$

Hence the trigonometric decomposition of $f_1(t)$:

$$\boxed{f_1(t) = \frac{1}{\pi} + \frac{1}{2}\sin t - \frac{2}{\pi}\sum_{p=1}^{p=\infty}\frac{1}{4p^2-1}\cos 2pt}$$

The coefficients for the complex spectrum appear as follows:

$$\begin{cases} C_0 = a_0 \\ C_n = \dfrac{1}{2}(a_n + jb_n) \\ C_{-n} = \dfrac{1}{2}(a_n - jb_n) \end{cases}$$

Thus: $$\boxed{\begin{array}{ll} C_0 = \dfrac{1}{\pi} & C_1 = C_{-1}^* = -j\dfrac{1}{4} \\ C_{2p} = \dfrac{-1}{\pi(4p^2-1)} & \end{array}}$$

From this we can plot the amplitude and phase spectra:

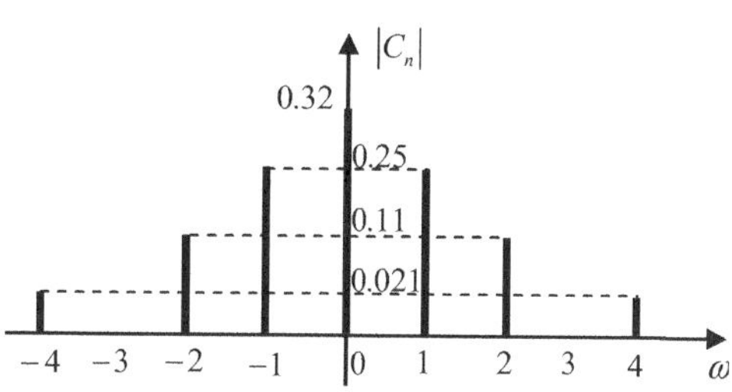

The complex spectrum is: $C_n = \frac{-j}{2} b_n$ and only includes imaginary components.

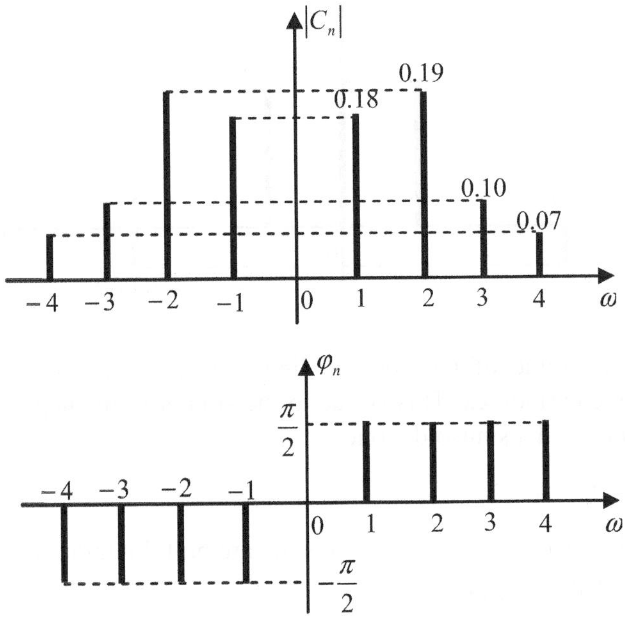

– Function $f_4(t)$:

Function $f_4(t)$ presents a rotational symmetry $f_4(t) = f_4(t \pm \frac{T}{2})$, this means we will only encounter odd harmonics.

Expressing this function in its analytical form on half a period suggests it would be better to use the exponential form of the Fourier series. After integration, we find the following result:

$$C_{2p+1} = \frac{j\left(1+e^{-\frac{\pi}{10}}\right)e^{-j(2p+1)\frac{\pi}{2}}}{\pi\left(\frac{1}{10}+j(2p+1)\right)}\sin(2p+1)\frac{\pi}{2}$$

This function can then be noted as follows:

$$f_4(t) = j\frac{1+e^{-\frac{\pi}{10}}}{\pi}\sum_{p=-\infty}^{p=+\infty}\frac{e^{-j(2p+1)\frac{\pi}{2}}}{\frac{1}{10}+j(2p+1)}\sin(2p+1)\frac{\pi}{2}e^{j(2p+1)t}$$

When the exponential time constant becomes infinite, C_{2p+1} becomes:

$$C_{2p+1} = \frac{\sin(2p+1)\frac{\pi}{2}}{(2p+1)\frac{\pi}{2}}e^{-j(2p+1)\frac{\pi}{2}}$$

This statement corresponds to the decomposition of a rectangular signal with a period of 2π and with a duty cycle of 2.

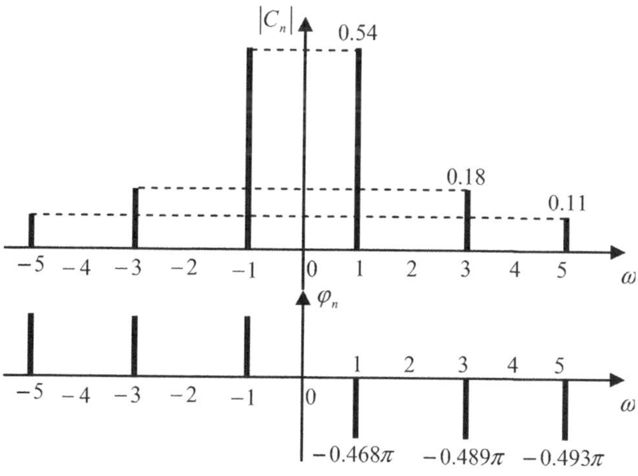

– Function $f_5(t)$:

Function $f_5(t)$ is even, coefficients b_n are null. Integration will take place over half a period, that period being $T = 4$ and the fundamental pulse stands at $\omega_0 = \frac{\pi}{2}$

$$a_n = \frac{4}{4}\int_0^1 \cos 200\pi t \cos n\frac{\pi}{2} t \, dt$$

This gives us:

$$a_n = \frac{2}{\pi}\frac{(-1)^{n+1} n}{(400+n)(400-n)} \text{ for } n \neq 400$$

If we repeat this operation for *n=400*, we obtain:

$$a_{400} = \frac{1}{2}$$

What's more: $a_0 = b_n = 0$

The Fourier decomposition will appear as follows:

$$f_5(t) = \frac{1}{2}\cos 200\pi t + {}_{n \neq 400}\sum_{n=1}^{n=\infty} \frac{2}{\pi}\frac{(-1)^{n+1} n}{(400+n)(400-n)}\cos n\frac{\pi}{2} t$$

The spectral lines will be centered around *n=±400* thus $\omega = \pm 200\pi$.

As the spectrum is real, we will trace $a_n = 2C_n$.

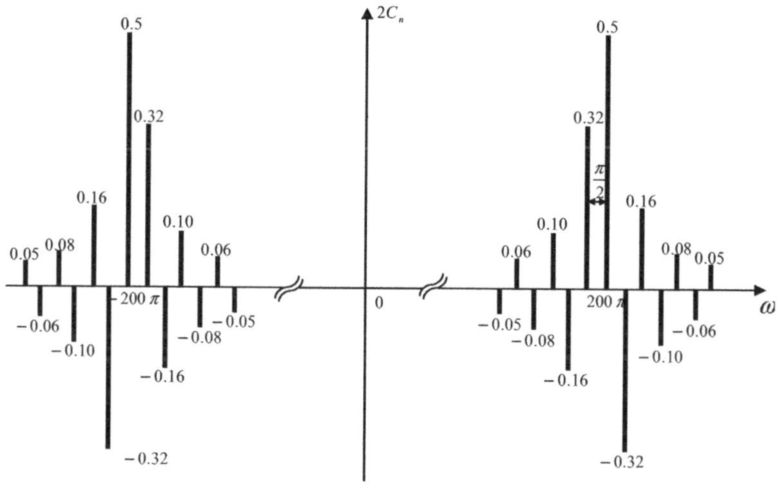

1.3.2. *Exercise 1.2*

Let us begin by proving that $f(t)$ is orthogonal to $\cos nt$ on interval $[0,2\pi]$:

$$I = \int_0^{2\pi} f(t)\cos nt\,dt = \int_0^{\pi} \cos nt\,dt + \int_{\pi}^{2\pi} -\cos nt\,dt$$

The integral is null, $f(t)$ is hence orthogonal to $\cos nt$.

This tells us that the Fourier decomposition will not include any cosine values.

Let us now show that the error function $f_e(t)$ is orthogonal to $\sin t$ on interval $[0,2\pi]$.

$$I' = \int_0^{2\pi} f_e(t)\sin t\,dt = \int_0^{\pi}\left(1-\frac{4}{\pi}\sin t\right)dt + \int_{\pi}^{2\pi}\left(-1-\frac{4}{\pi}\sin t\right)dt$$

The integral is null and $f_e(t)$ is hence orthogonal to $\sin t$.

The error function therefore cannot be expressed according to sin t. This means that $\frac{4}{\pi}\sin t$ is the fundamental value of the Fourier decomposition of $f(t)$.

1.3.3. *Exercise 1.3*

The energy of the function is noted as follows:

$$E=\int_{t_1}^{t_2} f^2(t)dt=\int_{t_1}^{t_2}\left(f_1(t)+f_2(t)\right)^2 dt=\int_{t_1}^{t_2} f_1^2(t)dt+\int_{t_1}^{t_2} f_2^2(t)dt+\int_{t_1}^{t_2} 2f_1(t)f_2(t)dt$$

The two first terms represent the energy of each function. The third is interaction energy between the two functions.

If both functions are orthogonal on interval $[t_1,t_2]$ then integral $\int_{t_1}^{t_2} f_1(t)f_2(t)dt=0$

The level of interaction energy is null. The resultant is immediately apparent.

1.3.4. *Exercice 1.4*

If function $f(t)$ is even, terms b_n must be null. The terms of the exponential decomposition appear as follow:

$$C_n=\frac{1}{2}(a_n-jb_n)=\frac{1}{2}a_n$$

This indicates they are real.

Similarly, when $f(t)$ is odd, terms a_n are null.

Hence: $C_n=\frac{1}{2}(a_n-jb_n)=-j\frac{1}{2}b_n$

The terms are therefore imaginary.

1.3.5. *Exercise 1.5*

Both functions $f(t)$ and $g(t)$ have the same Fourier decomposition except in their mean value. The difference between these two functions must be orthogonal to $\cos n\omega_0 t$ and $\sin n\omega_0 t$ regardless of n with $n \neq 0$.

We obtain $f(t) - g(t) = K = cte$, hence the solution: $f(t) = g(t) + K$

Two identical functions with different periods will have the same Fourier decomposition coefficients and only the frequencies of the various harmonics will differ.

$$T_1 = 2\pi\ s \Rightarrow \omega_1 = 1\ rd/s \Rightarrow \omega_n = n\ rd/s$$

$$T_2 = \pi\ ms \Rightarrow \omega_1 = 2.10^3\ rd/s \Rightarrow \omega_n = 2n.10^3\ rd/s$$

1.3.6. *Exercise 1.6*

– Function $f_1(t)$ is even and therefore $b_n = 0$. We have to integrate it into a half-period.

The period is $T = 4$ and the fundamental pulse is $\omega_0 = \frac{\pi}{2}$

$$a_0 = \frac{1}{2}\int_0^1 dt = \frac{1}{2} \quad a_n = \int_0^1 \cos n\frac{\pi}{2} t dt = \frac{\sin n\frac{\pi}{2}}{n\frac{\pi}{2}}$$

Thus: $f_1(t) = \frac{1}{2} + \sum_{n=1}^{n=\infty} \frac{\sin n\frac{\pi}{2}}{n\frac{\pi}{2}} \cos n\frac{\pi}{2} t$

All even harmonics are null. This is due to the rotational symmetry of the function that is being masked by the mean value.

We can rewrite the Fourier decomposition as:

$$f_1(t) = \frac{1}{2} + \sum_{p=0}^{p=\infty} \frac{\sin(2p+1)\frac{\pi}{2}}{(2p+1)\frac{\pi}{2}} \cos(2p+1)\frac{\pi}{2}t$$

The terms of the exponential decomposition then appear as follows:

$$C_{2p+1} = \frac{1}{2}\frac{\sin(2p+1)\frac{\pi}{2}}{(2p+1)\frac{\pi}{2}}$$

– Function $f_2(t)$ corresponds to the translation of function $f_1(t)$:

$$f_2(t) = f_1(t-1)$$

The coefficients in the exponential decomposition appear as expected, the amplitudes and the phases differ to $n\omega_0\tau = n\frac{\pi}{2}$:

$$C'_n = C_n e^{-jn\frac{\pi}{2}} \Rightarrow C'_{2p+1} = \frac{1}{2}\frac{\sin(2p+1)\frac{\pi}{2}}{(2p+1)\frac{\pi}{2}} e^{-j(2p+1)\frac{\pi}{2}}$$

$$C'_0 = \frac{1}{2}$$

– Function $f_3(t)$ is the result of translating and changing the mean value of $f_1(t)$:

$$f_3(t) = 2 + f_1(t-2)$$

These are the coefficients of the exponential decomposition:

$$C'_n = C_n e^{-jn\omega_0\tau} \Rightarrow C'_{2p+1} = \frac{1}{2}\frac{\sin(2p+1)\frac{\pi}{2}}{(2p+1)\frac{\pi}{2}} e^{-j(2p+1)\pi}$$

$$C'_0 = \frac{5}{2}$$

– Function $f_4(t)$ is the result of changing the mean value, amplitude, period and performing a time-shift on $f_1(t)$:

New amplitude: 2

Mean value: 0

Period: $T = 6 \Rightarrow \omega_0 = \frac{\pi}{3}$

Time-shift: $\tau = 2.5 \Rightarrow \omega_0\tau = \frac{5\pi}{6}$

$$C'_n = 2C_n e^{-jn\omega_0\tau} \Rightarrow C'_{2p+1} = \frac{\sin(2p+1)\frac{\pi}{3}}{(2p+1)\frac{\pi}{3}} e^{-j(2p+1)\frac{5\pi}{6}}$$

$$C'_0 = 0$$

1.3.7. *Exercise 1.7. Translation and composition of functions*

– Period is $T = 4\pi$, fundamental pulse is $\omega_0 = \frac{1}{2}$.

The exponential decomposition can be obtained from the following integral:

$$C_n = \frac{1}{4\pi}\int_0^{2\pi} \sin t e^{-jn\frac{t}{2}} dt$$

After calculations:

$$\begin{cases} C_{2p} = 0 \text{ for } p \neq \pm 1 & C_{2p+1} = \dfrac{2}{\pi} \dfrac{1}{4-(2p+1)^2} \\ C_2 = \dfrac{1}{4j} & C_{-2} = \dfrac{-1}{4j} \end{cases}$$

The exponential decomposition will therefore appear as follows:

$$f(t) = \frac{1}{2}\sin t + \sum_{p=-\infty}^{p=+\infty} \frac{1}{2\pi\left(1-\left(p+\frac{1}{2}\right)^2\right)} e^{j\left(p+\frac{1}{2}\right)t}$$

– $f_i(t)$ is an odd function. The decomposition contains only terms with sine. The terms of the exponential decomposition are easily obtained by performing a phase change of $-n\omega_0\tau$ that corresponds to time-shift $\tau = -\pi$, which would be $-(2p+1)\frac{\pi}{2}$.

We then have: $f_i(t) = -\dfrac{1}{2}\sin t + j\dfrac{2}{\pi}\displaystyle\sum_{p=-\infty}^{p=+\infty} \frac{(-1)^p}{4-(2p+1)^2} e^{j(2p+1)\frac{t}{2}}$

– The same method can be applied to function $f_i^{'}(t)$, the phase shift that we apply here has the opposite sign to function $f_i(t)$ which would be $(2p+1)\frac{\pi}{2}$

We then have: $f_i^{'}(t) = -\dfrac{1}{2}\sin t - j\dfrac{2}{\pi}\displaystyle\sum_{p=-\infty}^{p=+\infty} \frac{(-1)^p}{4-(2p+1)^2} e^{j(2p+1)\frac{t}{2}}$

– Function $f_1(t)$ can also be presented as follows:

$$f_1(t) = f(t - \frac{3\pi}{2}) - f(t + \frac{\pi}{2})$$

Starting with the decomposition of $f(t)$ and adding to that a phase change, we obtain the decompositions for $f(t-3\frac{\pi}{2})$ and $f(t+\frac{\pi}{2})$.

If we then subtract the decomposition for $f(t)$:

$$f_1(t) = -\frac{1+j}{\sqrt{2}} \sum_{p=-\infty}^{p=+\infty} \frac{e^{jp\frac{\pi}{2}}}{1-\left(p+\frac{1}{2}\right)^2} e^{j\left(p+\frac{1}{2}\right)t}$$

1.3.8. *Exercise 1.8. Time derivation of functions*

The derived function can be expressed as follows:

$$f^{'}(t+nT) = \lim_{\varepsilon \to 0} \frac{f(t+nT+\varepsilon) - f(t+nT)}{\varepsilon}$$

If the period for the function is T, the following can be stated:

$$f^{'}(t+nT) = \lim_{\varepsilon \to 0} \frac{f(t+\varepsilon) - f(t)}{\varepsilon} = f^{'}(t)$$

This means $f^{'}(t)$ is periodic and has a period of T.

$f^{'}(t)$ can then be decomposed into a Fourier series:

$$f^{'}(t) = \sum_{n=-\infty}^{n=+\infty} C_n^{'} e^{jn\omega_0 t}$$

With $C_n^{'} = \frac{1}{T}\int_0^T \frac{df(t)}{dt} e^{-jn\omega_0 t} dt$

After part by part integration we obtain: $C_n^{'} = jn\omega_0 C_n$

Hence: $f(t) = \sum_{n=-\infty}^{n=+\infty} C_n e^{jn\omega_0 t} \Rightarrow \frac{df(t)}{dt} = \sum_{n=-\infty}^{n=+\infty} jn\omega_0 C_n e^{jn\omega_0 t}$

1.3.9. *Exercise 1.9. Time integration of functions*

Consider the following periodic function $f(t) = \sum_{n=-\infty}^{n=+\infty} C_n e^{jn\omega_0 t}$ and its integral function $F(t)$, so that:

$$F(t) = \int f(t)dt$$

We can calculate $F(t+nT) - F(t) = [F(x)]_t^{t+nT} = \int_t^{t+nT} f(x)dx = nTC_0$

C_0 being the mean value of $f(t)$.

The integral function of $f(t)$ is only periodic with a period of T if the mean value for $f(t)$ is null.

$F(t)$ can then be decomposed into a Fourier series:

$$F(t) = \sum_{n=-\infty}^{n=+\infty} C'_n e^{jn\omega_0 t}$$

with: $C'_n = \frac{1}{T}\int_0^T F(t)e^{-jn\omega_0 t}dt$

After part-by-part integration: $C'_n = \frac{C_n}{jn\omega_0}$

1.3.10. *Exercise 1.10*

It only takes a few simple operations to get from $f(t)$ to $g(t)$:

– Remove mean value: $f_1(t) = f(t) - 1$

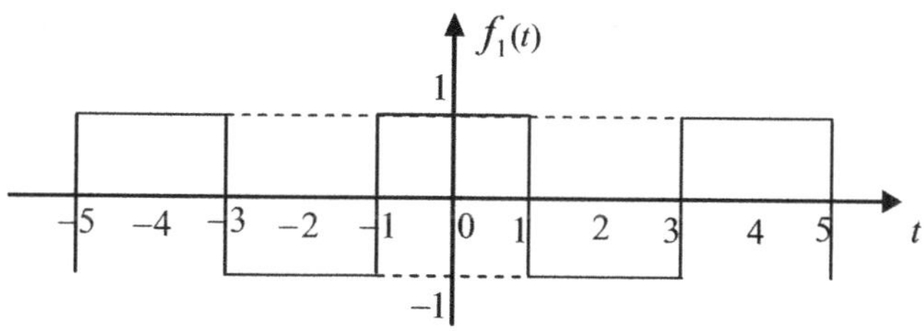

– Integrate: $f_2(t) = \int f_1(t)dt$

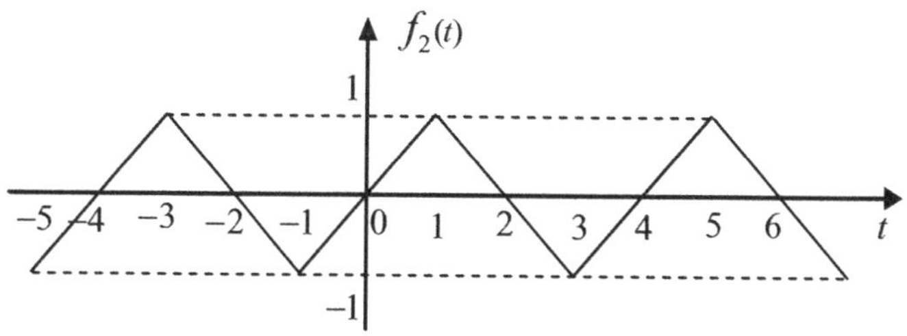

– Introduce mean value: $f_3(t) = f_2(t) + 1$

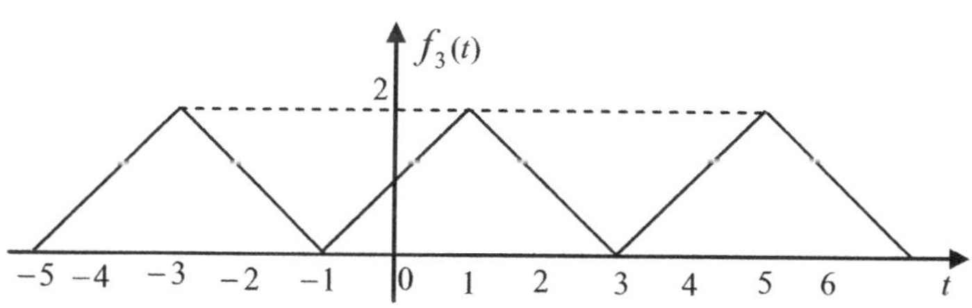

– Translate to the left: $g(t) = f_3(t+1)$

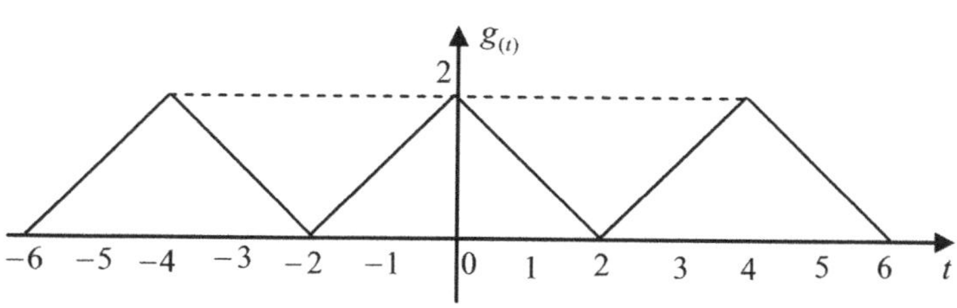

These various operations in the time-domain then translate very simply into the frequency domain.

$$f(t) = 1 + \sum_{n=-\infty}^{n=\infty} \frac{\sin n\frac{\pi}{2}}{n\frac{\pi}{2}} e^{jn\frac{\pi}{2}t} \Bigg|_{n\neq 0}$$

$$f_1(t) = \sum_{n=-\infty}^{n=\infty} \frac{\sin n\frac{\pi}{2}}{n\frac{\pi}{2}} e^{jn\frac{\pi}{2}t} \Bigg|_{n\neq 0}$$

$$f_2(t) = \sum_{n=-\infty}^{n=\infty} \frac{\sin n\frac{\pi}{2}}{j\left(n\frac{\pi}{2}\right)^2} e^{jn\frac{\pi}{2}t} \Bigg|_{n\neq 0}$$

$$f_3(t) = 1 + \sum_{n=-\infty}^{n=\infty} \frac{\sin n\frac{\pi}{2}}{j\left(n\frac{\pi}{2}\right)^2} e^{jn\frac{\pi}{2}t} \Bigg|_{n\neq 0}$$

$$g(t) = 1 + \sum_{n=-\infty}^{n=\infty} \frac{\sin n\frac{\pi}{2}}{j\left(n\frac{\pi}{2}\right)^2} e^{jn\frac{\pi}{2}} e^{jn\frac{\pi}{2}t} \Bigg|_{n\neq 0}$$

Note that the even terms are null. We obtain:

$$g(t) = 1 + \sum_{p=-\infty}^{p=\infty} \frac{4}{\pi^2 (2p+1)^2} e^{j(2p+1)\frac{\pi}{2}t}$$

1.3.11. *Exercise 1.11*

The impedance of a circuit as a sinusoidal system will appear:

$$Z(j\omega) = R + jL\omega + \frac{1}{jC\omega} = 0.01 + j\omega + \frac{16\pi^2}{j\omega}$$

We obtain the components of the Fourier decomposition of $i(t)$ by dividing those for $v(t)$ by $Z_{(j\omega)}$ set at the pulsation.

$$\begin{cases} v(t) = \sum_{n=-\infty}^{n=+\infty} C_n e^{jn\omega_0 t} \\ Z(j\omega) = 0.01 + jn\omega_0 + \frac{16\pi^2}{jn\omega_0} \\ i(t) = \sum_{n=-\infty}^{n=+\infty} \frac{C_n}{Z(jn\omega_0)} e^{jn\omega_0 t} = \sum_{n=-\infty n}^{n=+\infty} C'_n e^{jn\omega_0 t} \end{cases}$$

Both voltages have the same period $T = 1$ meaning $\omega_0 = 2\pi$.

The impedance at the different pulsations of the decomposition can be expressed as follows:

$$Z_{(jn\omega_0)} = 0.01 + j2\pi \frac{n^2 - 4}{n}$$

For $n=0$ which is the same as saying $\omega = 0$, the impedance becomes infinite. The continuous component, or mean value of the current will be null which is explained by the presence of a capacitor in the circuit.

The impedance is real and equal to 0.001 for $n = \pm 2$ or $\omega = \pm 4\pi = \pm 2\omega_0$ which is the resonant frequency of the RLC series circuit.

The Fourier decompositions for the different voltages are:

$$v_1(t) \Rightarrow C_n = -\frac{2}{\pi}\frac{1}{4n^2-1}$$

$$v_2(t) \Rightarrow \begin{cases} C_n = \dfrac{-1}{j2\pi n} \\ C_0 = \dfrac{1}{2} \end{cases}$$

This gives us the expression of the Fourier decompositions of the different currents and the spectral representation of the first three harmonics:

$$i_1(t) \Rightarrow \begin{cases} C_0^{'} = 0 \\ C_n^{'} = \dfrac{2}{\pi}\dfrac{1}{1-4n^2}\dfrac{1}{0.01+j2\pi\dfrac{n^2-4}{n}} \end{cases}$$

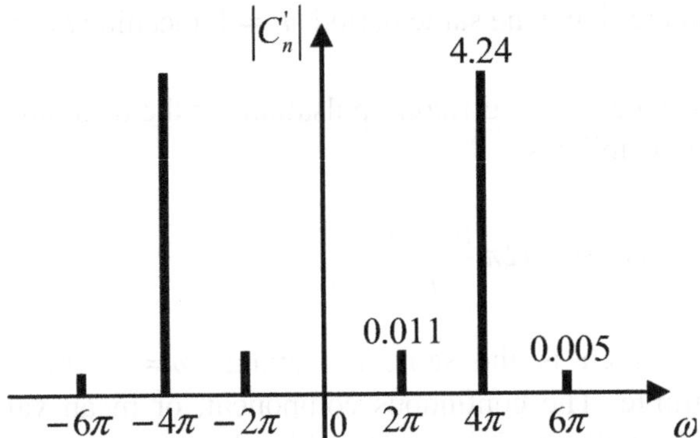

$$i_2(t) \Rightarrow \begin{cases} C_0^{'} = 0 \\ C_n^{'} = \dfrac{-1}{j2\pi n\left(0.01+j2\pi\dfrac{n^2-4}{n}\right)} \end{cases}$$

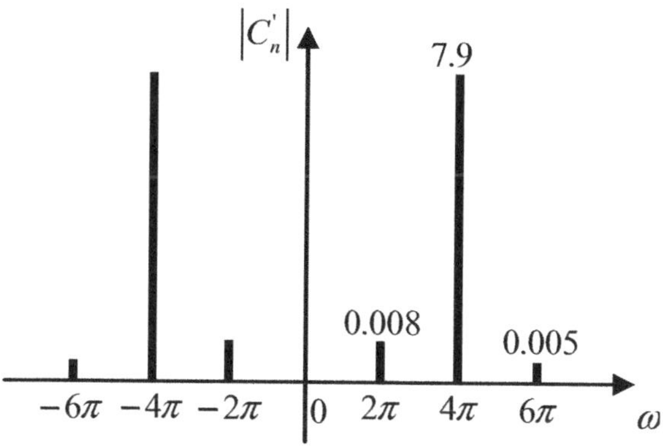

Spectral components other than $\omega = \pm 4\pi$ have a very low amplitude. This means the current is practically sinusoidal with a frequency double that of the input signal. This comes down to the selectivity of the resonant RLC series circuit.

2

Fourier Transform

2.1. Theoretical background

When a function is non-periodic, it becomes impossible to represent it as a Fourier series on a time interval $]-\infty,+\infty[$. The Fourier transform generalizes the frequency representation of a function, in a situation where said function is non-periodic.

2.1.1. *Fourier transform*

2.1.1.1. *Representation of any function* $]-\infty,+\infty[$

Let us consider any function $f(t)$ defined on $]-\infty,+\infty[$ so that $f(t)$ approaches 0 when $t \to \infty$

Let $f_T(t)$ be a periodic function of period T with $f_T(t) \equiv f(t)$ over one period.

We can state that: $f(t) = \lim\limits_{T\to\infty} f_T(t)$ [2.1]

$f_T(t)$ can be represented by a Fourier series.

$$f_T(t) = \sum_{n=-\infty}^{n=\infty} C_n e^{jn\omega_0 t}$$

$$C_n = \frac{1}{T}\int_{t_0-\frac{T}{2}}^{t_0+\frac{T}{2}} f_T(t)e^{-jn\omega_0 t}dt \quad \forall t_0 \qquad \text{with} \quad \omega_0 = \frac{2\pi}{T} \qquad [2.2]$$

We obtain a frequency representation of $f(t)$ by allowing T to approach infinity. In this case $C_n \to 0$ and $\omega_0 \to 0$, the spectrum approaches a continuous spectrum, the number of spectral lines becomes infinite but the amplitude of each line becomes infinitely small:

$$C(n\omega_0) = TC_n = \int_{t_0-\frac{T}{2}}^{t_0+\frac{T}{2}} f_T(t)e^{-jn\omega_0 t}dt \qquad [2.3]$$

We can then set the following: $f_T(t) = \frac{1}{2\pi}\sum_{n=-\infty}^{n=\infty} C(n\omega_0)e^{jn\omega_0 t}\omega_0$ [2.4]

Let us plot the real part of $C(n\omega_0)e^{jn\omega_0 t}$, this function only exists if $\omega = n\omega_0$.

The hatched area in Figure 2.1 is equal to $\omega_0 \Re e\left(C(n\omega_0)e^{jn\omega_0 t}\right)$. When $T \to \infty$ $\omega_0 \to 0$, the curve becomes continuous according to ω and the sum of the area of all the rectangles is equal to the area present between the curve and the pulsation axis.

$$\lim_{T\to\infty}\sum_{n=-\infty}^{n=\infty} \Re e\left(C(n\omega_0)e^{jn\omega_0 t}\right)\omega_0 = \int_{-\infty}^{+\infty} \Re e\left(C(\omega)e^{j\omega t}\right)d\omega$$

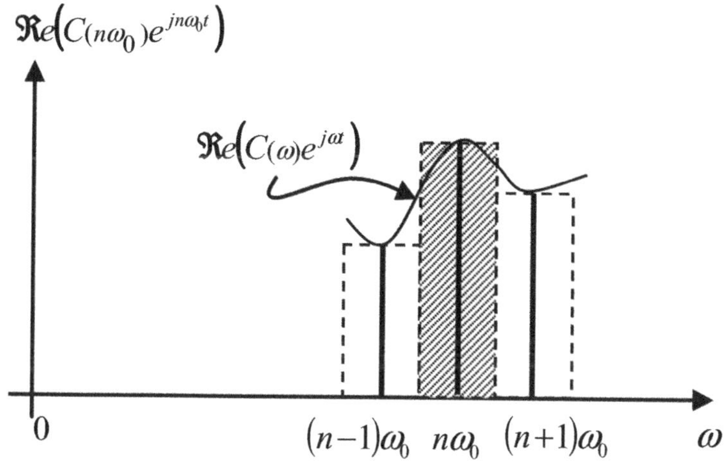

Figure 2.1. *Plot of* $\Re e\left(C(n\omega_0)e^{jn\omega_0 t}\right)$

We obtain identical results with the imaginary part of $C(n\omega_0)e^{jn\omega_0 t}$. It is then possible to group the results under the following form:

$$
\begin{aligned}
f(t) &= \frac{1}{2\pi}\int_{-\infty}^{+\infty} C(\omega)e^{j\omega t}\,d\omega \\
C(\omega) &= \int_{-\infty}^{+\infty} f(t)e^{-j\omega t}\,dt
\end{aligned}
\qquad [2.5]
$$

These equations define the direct and inverse Fourier transforms.

We can therefore represent $f(t)$ by a sum of continuous exponentials $e^{j\omega t}$ on interval $]-\infty,+\infty[$. The contribution of an infinitely small interval $d\omega$ is $\frac{1}{2\pi}C(\omega)d\omega$. $C(\omega)$ represents a density value:

$C(\omega)$ is the spectral density of function $f(t)$

We can also use a frequency variable: $\nu = \dfrac{\omega}{2\pi}$

The equations then become symmetric:

$$f(t) = \int_{-\infty}^{+\infty} C(\nu) e^{j2\pi\nu t} d\nu$$
$$C(\nu) = \int_{-\infty}^{+\infty} f(t) e^{-j2\pi\nu t} dt \qquad [2.6]$$

2.1.1.2. *Existence of the Fourier transform*

One way we can prove the existence of a Fourier transform is if $f(t)$ is absolutely integrable. If it is, there is necessarily a Fourier transform. However, this is not a necessary condition to the existence of a Fourier transform.

$\int_{-\infty}^{+\infty} |f(t)| dt$ exists and remains finite. $f(t) \in L_1$.

2.1.2. *Properties of the Fourier transform*

We will present the results but the developments will appear in the Bibliography.

2.1.2.1. *Linearity*

If $f_1(t) \xleftrightarrow{TF} F_1(j\omega)$, $f_2(t) \xleftrightarrow{TF} F_2(j\omega)$, $f_n(t) \xleftrightarrow{TF} F_n(j\omega)$

then $\sum_{i=1}^{i=n} f_i(t) \xleftrightarrow{TF} \sum_{i=1}^{i=n} F_i(j\omega)$

2.1.2.2. *Symmetry of the spectrum*

Let us consider a real function $f(t)$,

If $f(t) \xleftrightarrow{TF} F(j\omega)$ then, $F(j\omega) = F^*(-j\omega)$ [2.7]

Then $\left|F(j\omega)\right|$ is an even function and $Arg\left(F(j\omega)\right)$ is an odd function.

We will state that $F(j\omega)$ presents a Hermitian symmetry.

2.1.2.3. *Duality of the Fourier transform*

If $f(t) \xleftrightarrow{TF} F(j\omega)$, we demonstrate that $F(t) \xleftrightarrow{TF} 2\pi f(-j\omega)$ [2.8]

Particularly if $f(t)$ is even: $F(t) \xleftrightarrow{TF} 2\pi f(j\omega)$

Using these properties, we can easily determine a number of Fourier tranforms.

2.1.2.4. *Scale change*

If $f(t) \xleftrightarrow{TF} F(j\omega)$, then $\boxed{f(\alpha t) \xleftrightarrow{TF} \frac{1}{|\alpha|} F(j\frac{\omega}{\alpha})}$ with $\alpha \in R$ [2.9]

Creating a compression in time is the same as expanding the frequency scale.

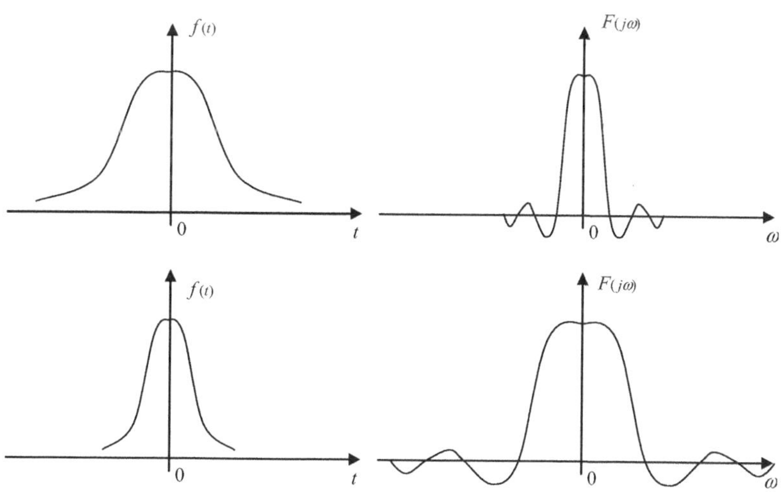

Figure 2.2. *Scale-change*

2.1.2.5. *Time and frequency scale translations*

$$\text{If } f(t) \overset{TF}{\longleftrightarrow} F(j\omega), \text{ then } \boxed{\begin{array}{l} f(t-t_0) \overset{TF}{\longleftrightarrow} F(j\omega)e^{-j\omega t_0} \\ f(t)e^{j\omega_0 t} \overset{TF}{\longleftrightarrow} F(j(\omega-\omega_0)) \end{array}} \qquad [2.10]$$

Frequency domain translations are common in the field of electronic applications. We can easily show that:

$$\boxed{\begin{array}{l} f(t)\cos\omega_0 t \overset{TF}{\longleftrightarrow} \frac{1}{2}\left[F(j(\omega-\omega_0)) + F(j(\omega+\omega_0))\right] \\ f(t)\sin\omega_0 t \overset{TF}{\longleftrightarrow} \frac{-j}{2}\left[F(j(\omega-\omega_0)) - F(j(\omega+\omega_0))\right] \end{array}}$$

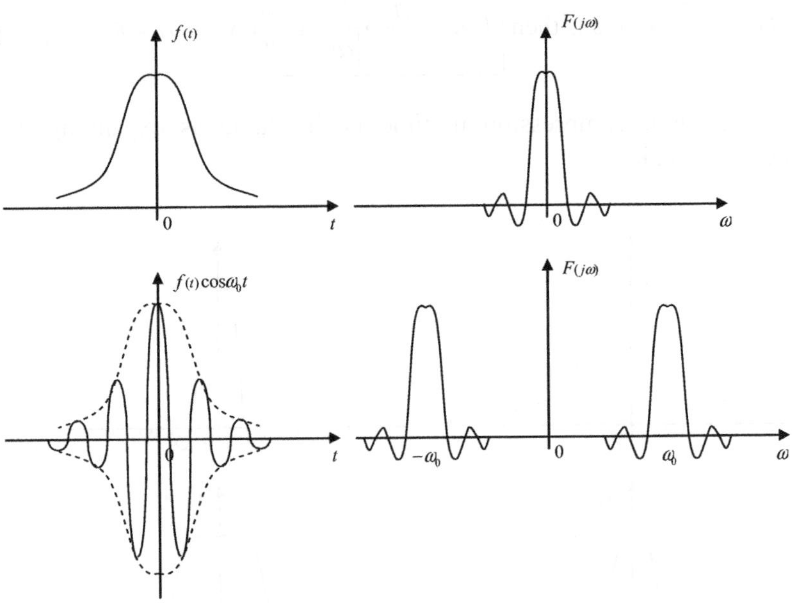

Figure 2.3. *Translation of the frequency scale*

2.1.2.6. *Integration and derivation in the time-domain*

If $f(t) \xleftrightarrow{TF} F(j\omega)$, then $\boxed{\dfrac{df(t)}{dt} \xleftrightarrow{TF} j\omega F(j\omega)}$ [2.11]

And in a more general sense $\dfrac{d^n f(t)}{dt^n} \xleftrightarrow{TF} (j\omega)^n F(j\omega)$

As such $\boxed{\int_{-\infty}^{t} f(x)dx \xleftrightarrow{TF} \dfrac{F(j\omega)}{j\omega}}$ [2.12]

The previous result is only valid if $\int_{-\infty}^{+\infty} f(x)dx = F(j\omega)\big|_{\omega=0} = 0$.

The opposite situation will be the centre of exercise 2.6.

2.1.2.7. *Derivation in the frequency domain*

If $f(t) \xleftrightarrow{TF} F(j\omega)$, then $\boxed{(-jt)^n f(t) \xleftrightarrow{TF} \dfrac{d^n F(j\omega)}{d\omega^n}}$ [2.13]

2.1.2.8. *Convolution*

We call convolution of functions $f_1(t)$ and $f_2(t)$ (see Chapter 5) the following expression:

$$f(t) = f_1(t) * f_2(t) = \int_{-\infty}^{+\infty} f_1(\tau) f_2(t-\tau) d\tau$$

If $f_1(t) \xleftrightarrow{TF} F_1(j\omega)$ and $f_2(t) \xleftrightarrow{TF} F_2(j\omega)$, then:

$$\begin{cases} f_1(t) * f_2(t) \xleftrightarrow{TF} F_1(j\omega) F_2(j\omega) \\ f_1(t) f_2(t) \xleftrightarrow{TF} \dfrac{1}{2\pi} F_1(j\omega) * F_2(j\omega) \end{cases} \qquad [2.14]$$

2.1.3. *Singular functions*

A singular function is one that includes points for which the derived forms are not defined as functions.

2.1.3.1. *Dirac delta function*

Consider the following function $f_\tau(t)$ (Figure 2.4).

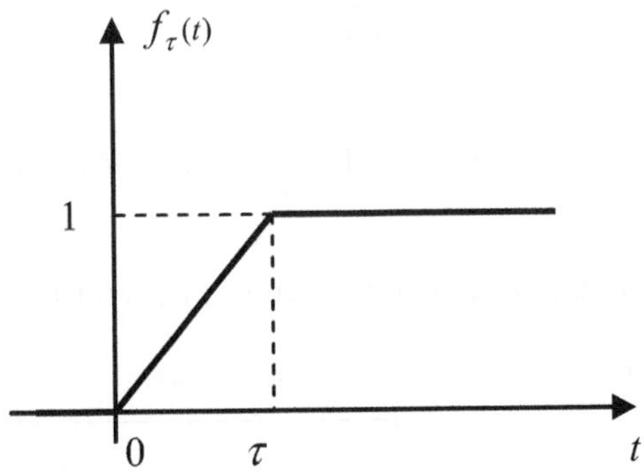

Figure 2.4. *Ramp function*

When τ approaches 0, $f_\tau(t) \to u(t)$ unit step function defined by:

$$u(t) = \begin{cases} 1 \text{ for } t \geq 0 \\ 0 \text{ for } t < 0 \end{cases}$$

The derivative $f_\tau'(t)$ is a rectangular impulse with a width of τ and an amplitude of $\dfrac{1}{\tau}$.

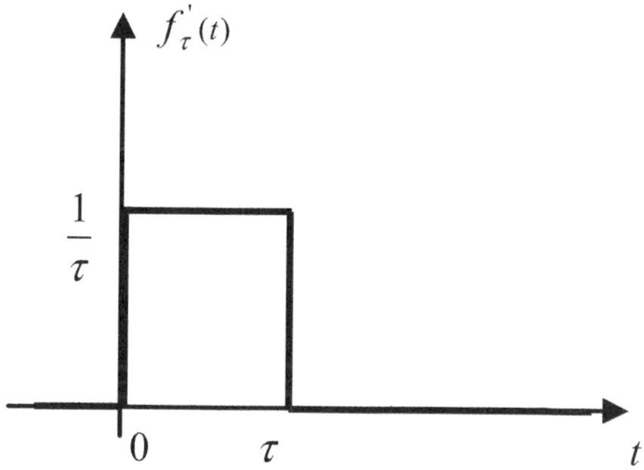

Figure 2.5. *Derivative of the ramp function*

When $\tau \to 0$, $f_\tau'(t)$ approaches a null impulse width, infinite amplitude but with a constant amount – equal to 1 – of surface under its curve.

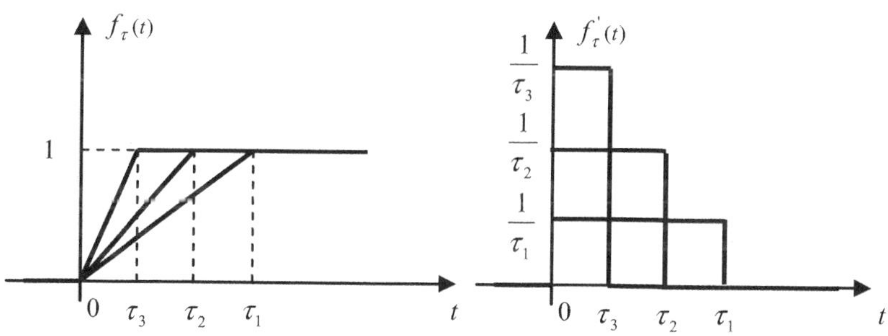

Figure 2.6. *Evolution following τ*

We define a new "function" $\delta(t)$ called Dirac impulse unit (or Dirac impulse) with a null width, infinite amplitude for $t = 0$, null on all other aspects and with a surface equal to 1. A more in-depth definition is provided in Chapter 6.

$$\begin{cases} \delta(t) = 0 \quad \text{for } t \neq 0 \\ \int_{-\infty}^{+\infty} \delta(t) dt = 1 \end{cases} \qquad [2.15]$$

This definition is unrelated to the form of the impulse used for approaching the limit. We could also have used the following forms (see exercises 2.3, 2.4 and 2.5):

– Gaussian impulse:

$$\delta(t) = \lim_{\tau \to 0} \left(\frac{1}{\tau} e^{-\pi \frac{t^2}{\tau^2}} \right)$$

– Triangular impulse:

$$\delta(t) = \begin{cases} \lim_{\tau \to 0} \left(\frac{1}{\tau} \left(1 - \frac{|t|}{\tau} \right) \right) & \text{for } |t| < \tau \\ 0 & \text{for } |t| > \tau \end{cases}$$

– Exponential impulsions:

$$\delta(t) = \lim_{\tau \to 0} \left(\frac{1}{2\tau} e^{-\frac{|t|}{\tau}} \right)$$

$$\delta(t) = \lim_{\tau \to 0} \left(\frac{1}{\tau} e^{-\frac{t}{\tau}} \right) u(t)$$

– Sine cardinal (Sinc) impulses:

$$\delta(t) = \lim_{a \to \infty} \left(\frac{a}{\pi} \frac{\sin at}{at} \right)$$

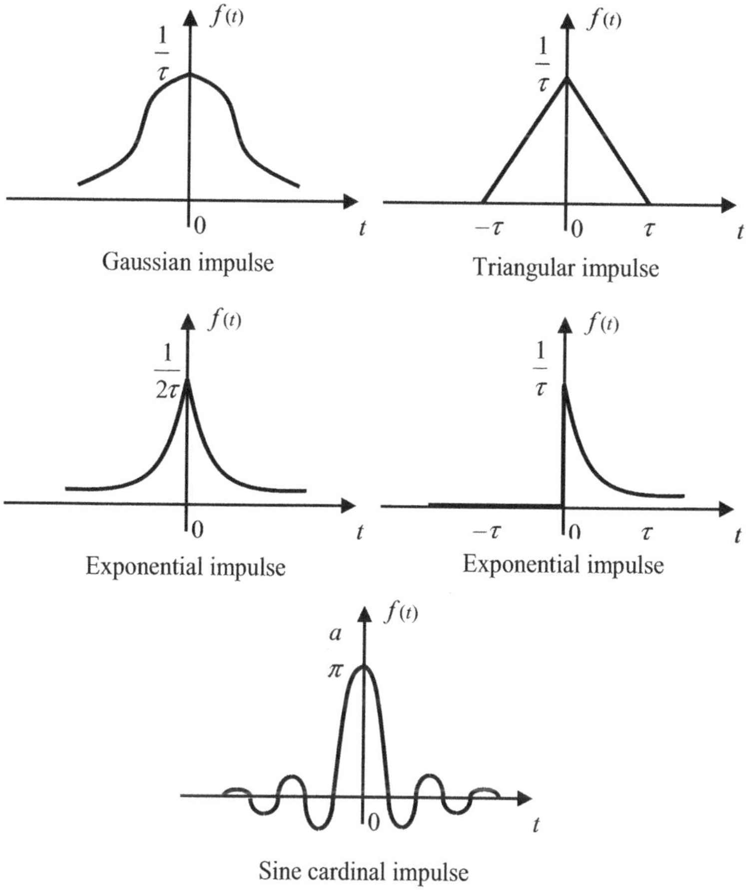

Figure 2.7. *Different forms of impulse*

Note that this definition relies on an extension of the classic notion of a function. This book will not be covering more in-depth uses for distribution theories. If readers would like to know more on this subject, see the books cited in the Bibliography [ROD 93, DEM 96, LES 12]. A concise definition will nonetheless be presented in Chapter 6.

2.1.3.2. *Derivatives for discontinuous functions*

"Function" $\delta(t)$ is an impulse with a weight of 1, $a\delta(t)$ is an impulse with a weight of a. The weight of the impulse is defined by:

$$\int_{-\infty}^{+\infty} a\delta(t) = a$$

Using "function" $\delta(t)$ allows us to calculate the derivatives of functions presenting a discontinuity (Figure 2.8).

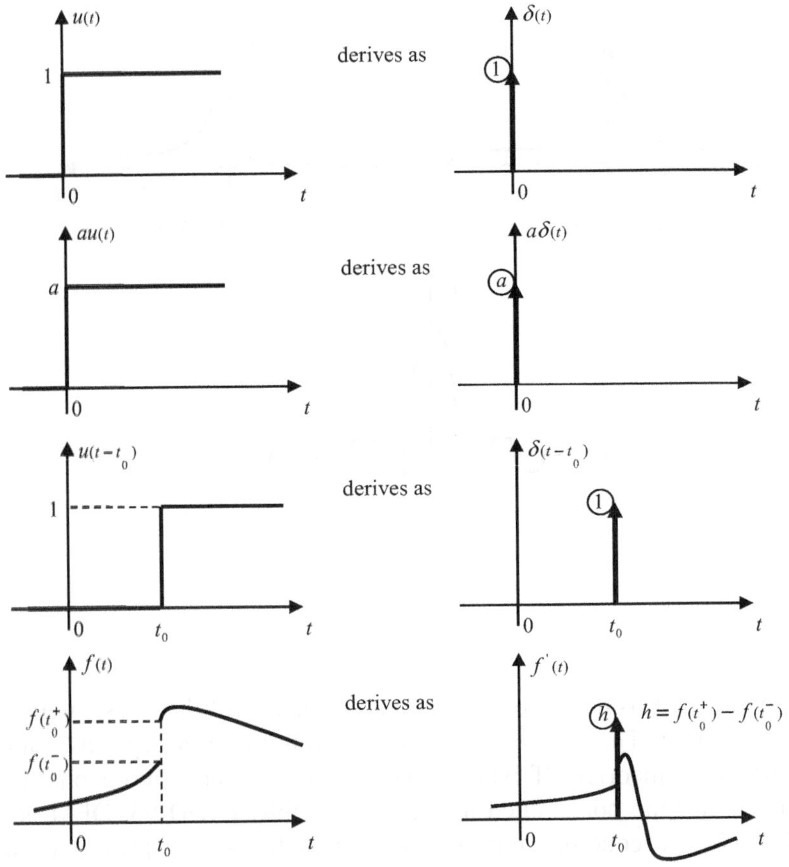

Figure 2.8. *Derivatives of discontinuous functions*

2.1.3.3. *Sampling property of* $\delta(t)$

Multiplying a function with an impulse $\delta(t)$ results in an impulse with a weight equal to the value of the function at the time of the localization of the impulse:

$$\int_{-\infty}^{+\infty} f(t)\delta(t-t_0) = f(t_0)$$

Thus $$\begin{cases} f(t)\delta(t) = f(0)\delta(t) \\ f(t)\delta(t-t_0) = f(t_0)\delta(t-t_0) \end{cases} \quad [2.16]$$

2.1.4. *Fourier transform of common functions*

The following table presents a number of Fourier transforms that are likely to appear in applied fields.

$f(t)$	$F(j\omega)$
$e^{-\alpha t}u(t)$	$\dfrac{1}{\alpha + j\omega}$
$te^{-\alpha t}u(t)$	$\dfrac{1}{(\alpha + j\omega)^2}$
$\delta(t)$	1
1	$2\pi\delta(\omega)$
$u(t)$	$\dfrac{1}{j\omega} + \pi\delta(\omega)$
$\cos\omega_0 t$	$\pi\left(\delta(\omega-\omega_0)+\delta(\omega+\omega_0)\right)$
$\cos\omega_0 t.u(t)$	$\dfrac{\pi}{2}\left(\delta(\omega-\omega_0)+\delta(\omega+\omega_0)\right)+\dfrac{j\omega}{\omega_0^2-\omega^2}$
$\sin\omega_0 t$	$-j\pi\left(\delta(\omega-\omega_0)-\delta(\omega+\omega_0)\right)$

<table>
<tr><td>$\sin \omega_0 t \, u(t)$</td><td>$-j\frac{\pi}{2}\left(\delta(\omega-\omega_0)-\delta(\omega+\omega_0)\right)+\frac{\omega_0}{\omega_0^2-\omega^2}$</td></tr>
<tr><td>$e^{-\alpha|t|}$</td><td>$\frac{2\alpha}{\alpha^2+\omega^2}$</td></tr>
<tr><td>$f(t)$</td><td>$F(j\omega)$</td></tr>
<tr><td>$e^{-\frac{t^2}{2\tau^2}}$</td><td>$e^{-\frac{t^2}{2\tau^2}}$</td></tr>
<tr><td>$\begin{cases} 1-\frac{|t|}{\tau} & |t|<\tau \\ 0 & |t>\tau| \end{cases}$</td><td>$\tau\left(\frac{\sin\frac{\omega\tau}{2}}{\frac{\omega\tau}{2}}\right)^2=\tau\sin c^2\frac{\omega\tau}{2}$</td></tr>
<tr><td>$Cr_\tau(t)$
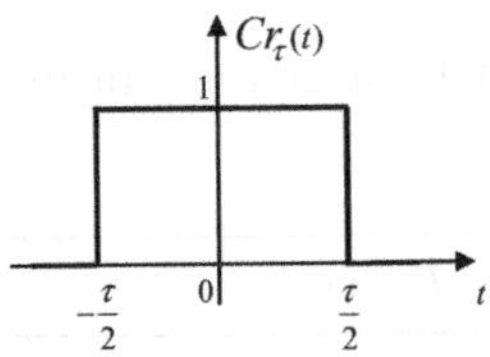
</td><td>$\tau\frac{\sin\frac{\omega\tau}{2}}{\frac{\omega\tau}{2}}=\tau\sin c\frac{\omega\tau}{2}$</td></tr>
<tr><td>$\frac{\omega_0}{2\pi}\frac{\sin\frac{\omega_0 t}{2}}{\frac{\omega_0 t}{2}}$</td><td>$Cr_{\omega_0}(\omega)$
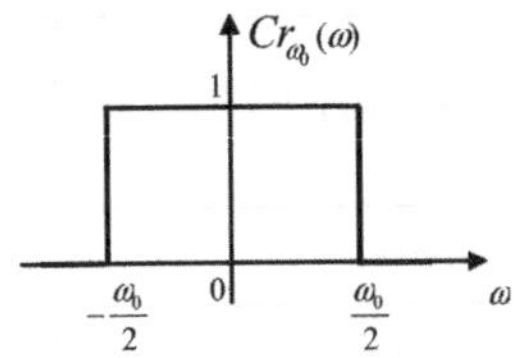
</td></tr>
<tr><td>$\delta_T(t)$
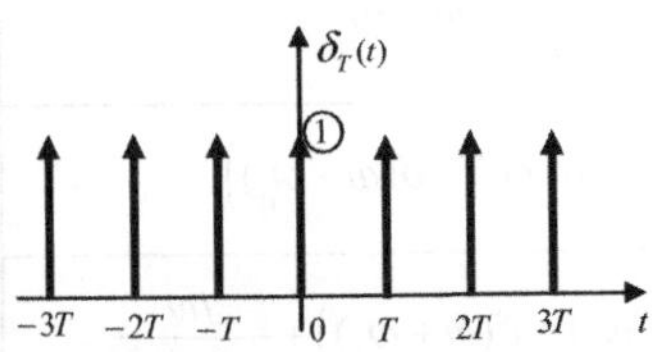
</td><td>$\omega_0\delta_{\omega_0}(\omega)$
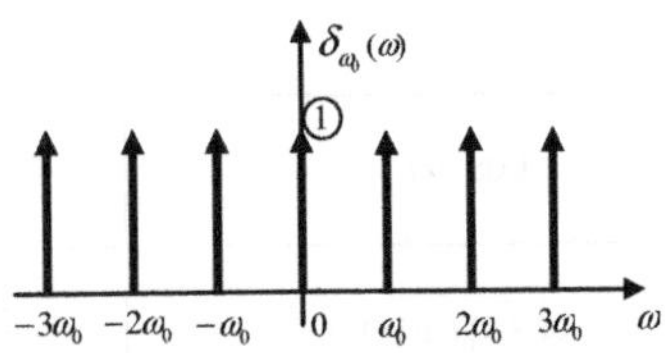
</td></tr>
</table>

Table 2.1. *Fourier transform of common function*

2.1.5. *Calculating Fourier transforms using the Dirac impulse method*

Consider a function made up of a succession of line segments with no discontinuation. This function can serve as an approximation of a non-linear function.

By deriving twice, we obtain a series of Dirac impulses located at the instants of change in slope as shown in Figure 2.9.

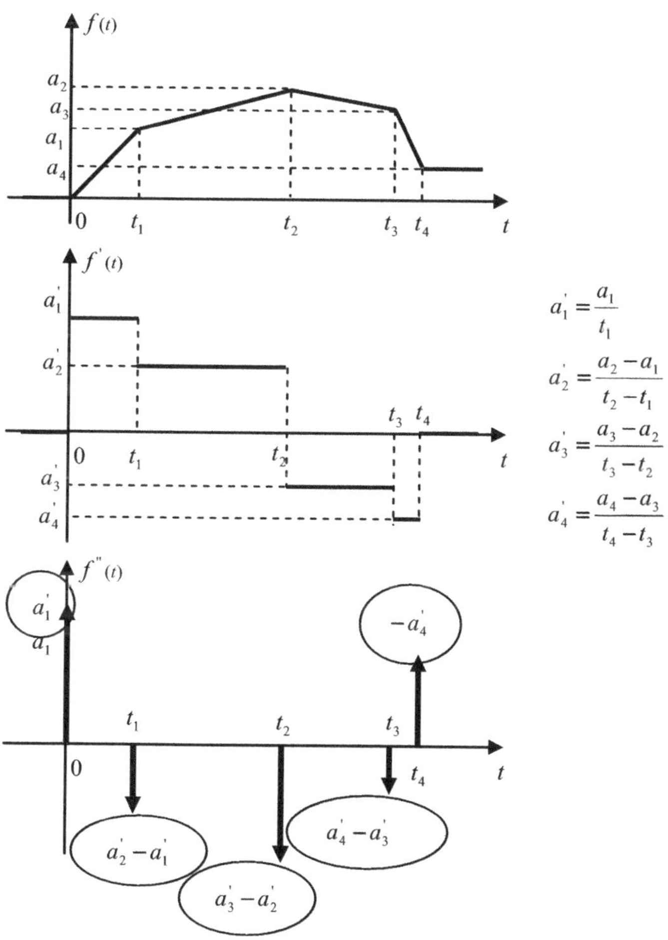

Figure 2.9. *Linear function defined in parts and its derivatives*

If we set h_i the weight of the impulse at moment t_i, the second derivative would appear as follows:

$$f''(t) = \sum_i h_i \delta(t_i)$$

Its Fourier transform would then be:

$$f''(t) = \sum_i h_i \delta(t_i) \xleftrightarrow{TF} \sum_i h_i e^{-j\omega t_i}$$

$$\text{If } f(t) \xleftrightarrow{TF} F(j\omega), \text{ then } f''(t) \xleftrightarrow{TF} -\omega^2 F(j\omega)$$

$$\text{Thus: } F(j\omega) = \frac{-1}{\omega^2} \sum_i h_i e^{-j\omega t_i} \qquad [2.17]$$

Examples of this are presented in exercise 2.2.

2.1.6. *Fourier transform of periodic functions*

Let us consider a periodic function $f(t)$ with a period of T and C_n the coefficient of the exponential Fourier series. The Fourier transform has the following form:

$$f(t) \xleftrightarrow{TF} 2\pi \sum_{n=-\infty}^{n=+\infty} C_n \delta(\omega - n\omega_0) \text{ with } \omega_0 = \frac{2\pi}{T} \qquad [2.18]$$

2.1.7. *Energy density*

The energy contained in a non-periodic signal appears as follows:

$$E = \int_{-\infty}^{+\infty} f^2(t) dt$$

We demonstrate that $\int_{-\infty}^{+\infty} f^2(t)dt = \frac{1}{2\pi}\int_{-\infty}^{+\infty} F(j\omega)F(-j\omega)d\omega$ if $f(t)$ is real.

In the case of a real function, we set the following:

$$\int_{-\infty}^{+\infty} f^2(t)dt = \frac{1}{2\pi}\int_{-\infty}^{+\infty} \left|F(j\omega)\right|^2 d\omega$$

Spectral density is what we would call the function below:

$$S(\omega) = \frac{1}{2\pi}\left|F(j\omega)\right|^2 \qquad [2.19]$$

The energy contained in the signal is then written:

$$E = \int_{-\infty}^{+\infty} S(\omega)d\omega$$

2.1.8. *Upper limits to the Fourier transform*

Let us consider a function $f(t)$, we can define for it three characteristic values:

– its content: $C = \int_{-\infty}^{+\infty} \left|f(t)\right| dt$

– its variation: $V = \int_{-\infty}^{+\infty} \left|\frac{df(t)}{dt}\right| dt$

– its ripple: $O = \int_{-\infty}^{+\infty} \left|\frac{d^2 f(t)}{dt^2}\right| dt$

If $F(j\omega)$ is the Fourier transform of $f(t)$, we can demonstrate that:

$$\begin{cases} \left|F(j\omega)\right| \leq C \\ \left|F(j\omega)\right| \leq \dfrac{V}{|\omega|} \\ \left|F(j\omega)\right| \leq \dfrac{O}{\omega^2} \end{cases}$$

These relations allow us to find the upper limits of $\left|F(j\omega)\right|$ (see exercise 2.15).

2.2. Exercises

The exercises presented in this chapter come from [LAT 66].

2.2.1. *Exercise 2.1*

Demonstrate that the Fourier transform can be written in the following form:

$$F(j\omega) = \int_{-\infty}^{+\infty} f(t)\cos\omega t\,dt - j\int_{-\infty}^{+\infty} f(t)\sin\omega t\,dt$$

From this, explain how if $f(t)$ is even: $F(j\omega) = 2\int_{0}^{+\infty} f(t)\cos\omega t\,dt$

And that if $f(t)$ is odd: $F(j\omega) = -2j\int_{0}^{+\infty} f(t)\sin\omega t\,dt$

From this, reach the following results:

$f(t)$	$F(j\omega)$
Real and even	Real and even
Real and odd	Imaginary and odd
Imaginary and even	Imaginary and even
Imaginary and odd	Real and odd
Complex and odd	Complex and even
Complex and odd	Complex and odd

2.2.2. *Exercise 2.2*

Determine the Fourier transforms of the following functions. Find these results using the Dirac impulse method where possible.

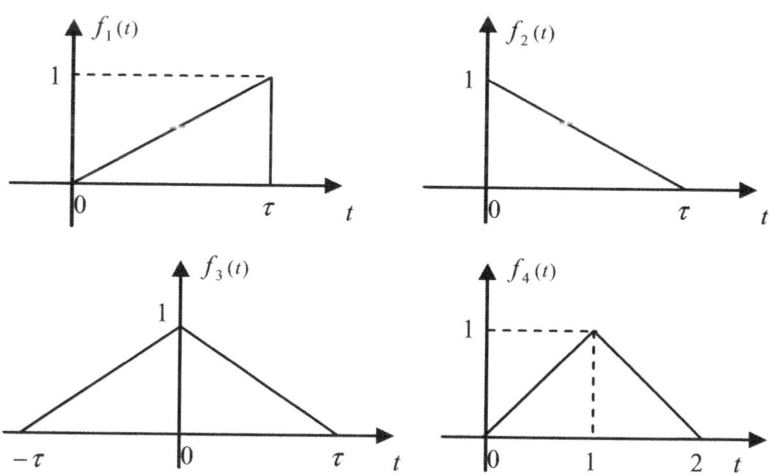

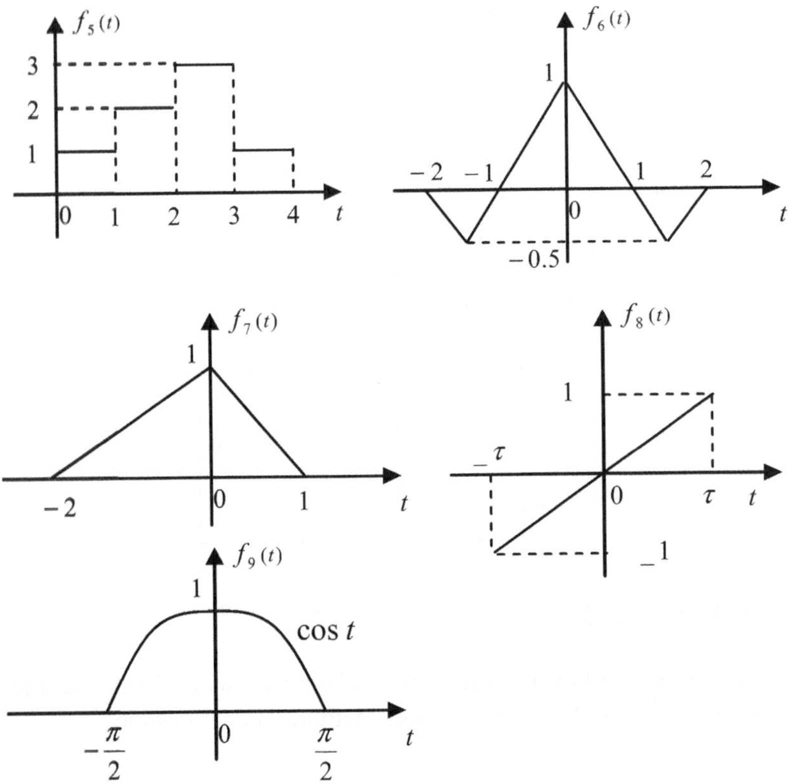

2.2.3. *Exercise 2.3*

If $\delta(t)$ is the Dirac impulse, demonstrate that:

$$\delta(t) = \frac{1}{2\pi}\int_{-\infty}^{+\infty} e^{j\omega t}\,d\omega$$

and

$$\delta(t) = \frac{1}{\pi}\int_{0}^{+\infty} \cos\omega t\,d\omega$$

We will use the Fourier transform of $\delta(t)$.

2.2.4. *Exercise 2.4*

Determine the Fourier transforms of the following functions:

$$f_1(t) = \begin{cases} \dfrac{1}{\tau}\left(1 - \dfrac{|t|}{\tau}\right) & \text{when } |t| \le \tau \\ 0 & \text{when } |t| > \tau \end{cases}$$

$$f_2(t) = \frac{1}{\tau} e^{-\frac{t}{\tau}} u(t)$$

$$f_3(t) = \frac{1}{\tau} e^{\frac{t}{\tau}} u(-t)$$

$$f_4(t) = \frac{1}{2\tau} e^{-\frac{|t|}{\tau}}$$

What limits do these Fourier transforms approach when τ approaches 0? Explain these results.

2.2.5. *Exercise 2.5*

Demonstrate that:

$$\delta(t) = \lim_{\alpha \to \infty} \left[\frac{\alpha}{\pi} \left(\frac{\sin \alpha t}{\alpha t} \right)^2 \right]$$

We can use the following result:

$$\int_{-\infty}^{+\infty} \left(\frac{\sin x}{x} \right)^2 dx = \pi$$

2.2.6. *Exercise 2.6*

Demonstrate that if $f(t)$ is not entirely integrable:

$$\int_{-\infty}^{+\infty} f(\tau) d\tau = F(j\omega)\big|_{\omega=0} \neq 0 \quad \text{with} \quad f(t) \overset{TF}{\leftrightarrow} F(j\omega)$$

Then the integration theorem becomes:

$$\int_{-\infty}^{t} f(\tau)d\tau \xleftrightarrow{TF} \frac{F(j\omega)}{j\omega} + \pi F(j\omega)\big|_{\omega=0}\delta(\omega)$$

We set $\int_{-\infty}^{t} f(\tau)d\tau$ as the convolution of $f(t)$ and $u(t)$.

2.2.7. *Exercise 2.7*

Let us consider a rectangular function $Cr_\tau(t)$:

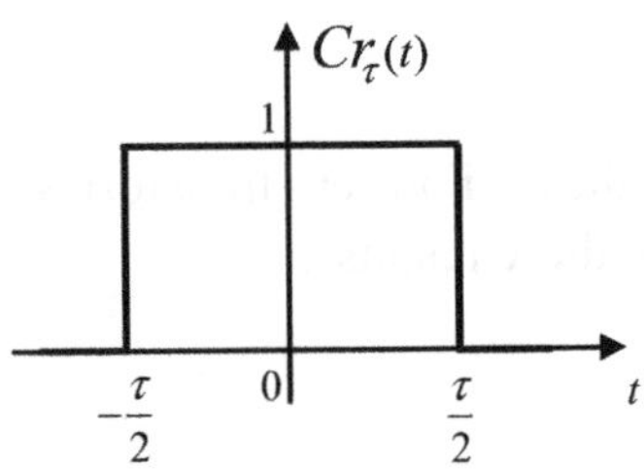

Determine the Fourier transforms of the following functions:

$$f_1(t) = \cos\omega_0 t$$

$$f_2(t) = \cos\omega_0 t Cr_T(t) \qquad \text{with } T = \frac{2\pi}{\omega_0}$$

$$f_3(t) = \cos\omega_0 t Cr_{20T}(t)$$

$$f_4(t) = \cos\omega_0 t u(t)$$

Plot the spectra and compare the results.

2.2.8. *Exercise 2.8*

Consider functions $F(j\omega)$ of module $|F(j\omega)|$ along with the following argument: $\phi(\omega) = Arg\left(F(j\omega)\right)$. Determine functions $f(t)$

that have $F(j\omega)$ as a Fourier transform in the following cases. We can use the reciprocity theorem:

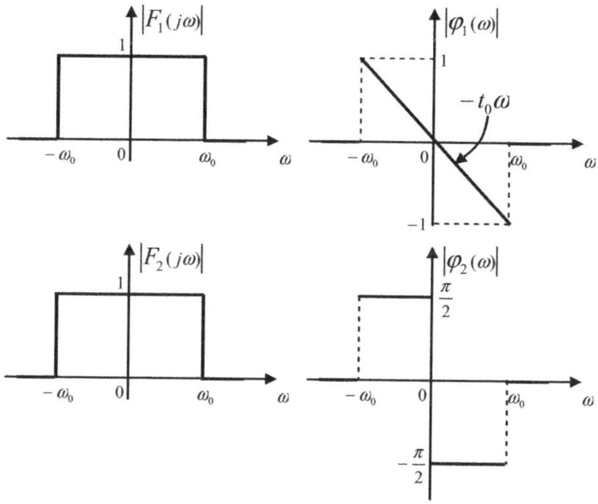

2.2.9. *Exercise 2.9*

Determine the Fourier transforms of the following functions:

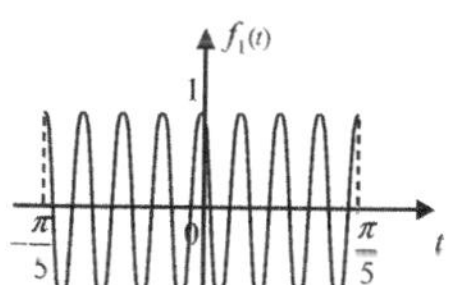

$$f_1(t)=\begin{cases}\cos 20t & \text{for } t\in\left[-\frac{\pi}{5},+\frac{\pi}{5}\right]\\ 0 & \text{for } |t|>\frac{\pi}{5}\end{cases}$$

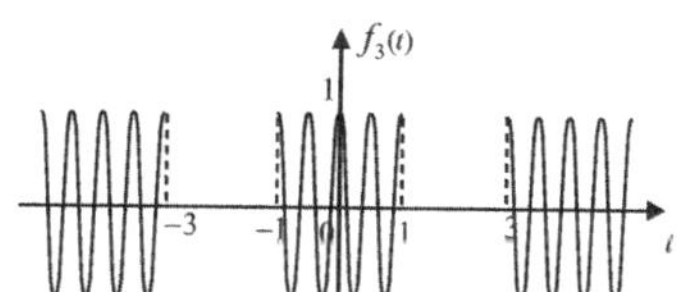

$$f_3(t)=\begin{cases}\cos 200\pi t & \text{for } t\in[-1+4n,1+4n]\\ 0 & \text{for } t\in[1+4n,3+4n]\end{cases}\quad \text{for } n\in Z$$

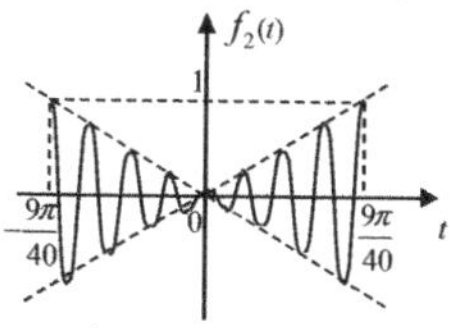

$$f_2(t)=\begin{cases}-\frac{40}{9\pi}t\cos 20t & \text{for } t\in\left[-\frac{9\pi}{40},+\frac{9\pi}{40}\right]\\ 0 & \text{for } |t|>\frac{9\pi}{40}\end{cases}$$

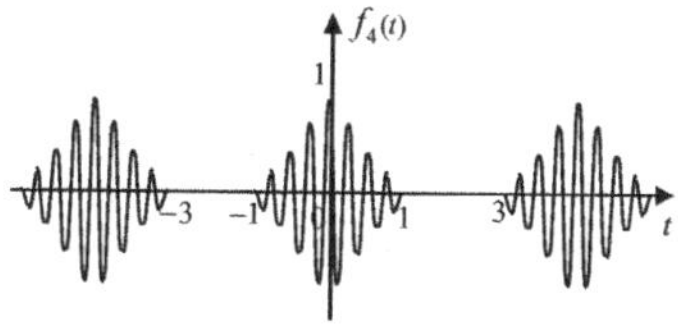

$$f_4(t)=\begin{cases}(1-|t|)\cos 200\pi t & \text{for } t\in[-1+4n,1+4n]\\ 0 & \text{for } t\in[1+4n,3+4n]\end{cases}\quad \text{for } n\in Z$$

We'll use the fact that all these functions are written with the following form: $f(t)\cos\omega_0 t$

2.2.10. *Exercise 2.10*

Using frequency-domain translations determine functions $f(t)$ that allow the following Fourier transforms:

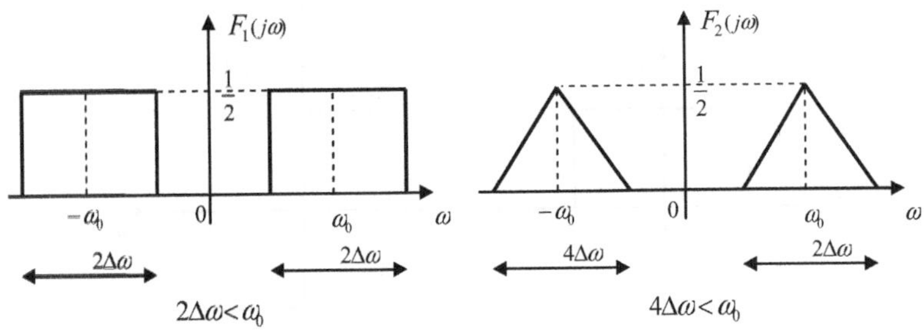

2.2.11. *Exercise 2.11*

If $f(t) \xleftrightarrow{TF} F(j\omega)$, demonstrate that:

$$f(t+T) + f(t-T) \xleftrightarrow{TF} 2\cos\omega T \, F(j\omega)$$

Use this result to calculate the Fourier transforms for the following functions:

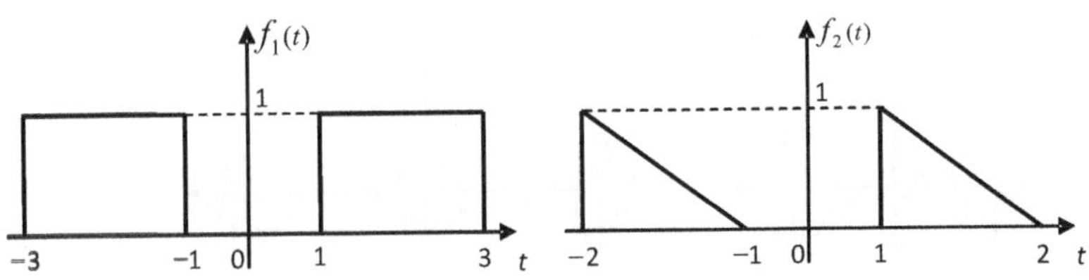

2.2.12. *Exercise 2.12*

Consider the following function $f(t)$:

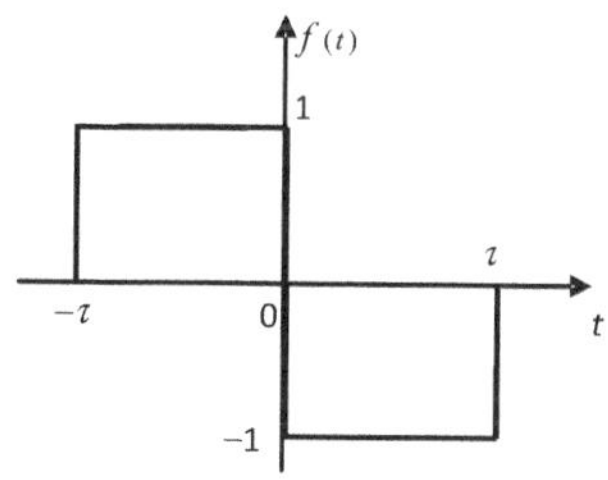

Determine its Fourier transform in the following different ways:

– direct calculation;

– using integration properties and the results table;

– using derivation properties and the results table;

– using time translation properties and the results table.

2.2.13. *Exercise 2.13*

We define the Dirac comb "function" as shown below:

$$\delta_T(t) = \sum_{n=-\infty}^{n=+\infty} \delta(t - nT)$$

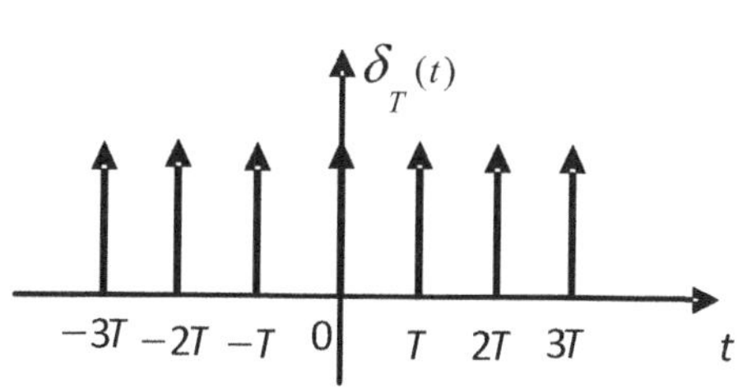

Determine the Fourier transform for this "function".

2.2.14. *Exercise 2.14*

Let us now consider that $f(t)$ is limited in time with a Fourier transform of $F(j\omega)$. From this function, let us form another function we will call $f_n(t)$ defined by:

$$f_n(t) = \sum_{p=-n}^{p=+n} f(t-pT) \text{ with } T > \text{temporal occupation of } f(t)$$

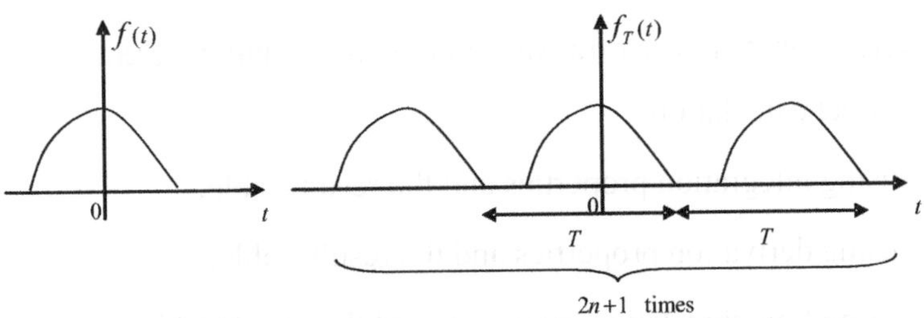

Demonstrate that $f_n(t) \overset{TF}{\longleftrightarrow} F(j\omega)\dfrac{\sin(2n+1)\dfrac{\omega T}{2}}{\sin\dfrac{\omega T}{2}}$

What happens to this result when $n \to \infty$? Explain this result.

We refer to this result as: $\displaystyle\sum_{k=-n}^{k=+n} e^{jkx} = \frac{\sin(2n+1)\dfrac{x}{2}}{\sin\dfrac{x}{2}}$

2.2.15. *Exercise 2.15*

Consider the following function:

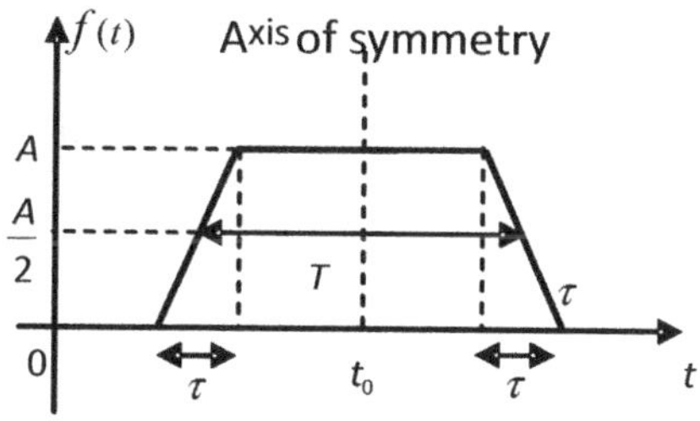

– determine $F(j\omega)$ the Fourier transform of $f(t)$;

– determine content C, variation V and ripple O of $f(t)$;

– check that these values allow us to find an upper limit to $|F(j\omega)|$. Plot these different limits.

2.2.16. *Exercise 2.16*

We define moment of order n of a function as:

$$m_n = \int_{-\infty}^{+\infty} t^n f(t)dt$$

Demonstrate that if $f(t) \xleftrightarrow{TF} F(j\omega)$ then:

$$m_n = j^n \left. \frac{d^n F(j\omega)}{d\omega^n} \right|_{\omega=0}$$

From this, establish that the decomposition into a Taylor series $F_{(j\omega)}$ appears as follows:

$$F(j\omega) = \sum_{n=0}^{n=\infty} (-j)^n m_n \frac{\omega^n}{n!}$$

2.2.17. *Exercise 2.17*

Consider function $f(t)$, with no impulse at its origin, null when $t<0$. We say of this function that it is causal.

Let us consider $f(t) \xleftrightarrow{TF} F(j\omega)$ and write

$$F(j\omega) = R(j\omega) + jX(j\omega).$$

From this, show that $\begin{cases} R(j\omega) = \dfrac{1}{\pi}\displaystyle\int_{-\infty}^{+\infty} \dfrac{X(x)}{\omega - x}\,dx \\ X(j\omega) = -\dfrac{1}{\pi}\displaystyle\int_{-\infty}^{+\infty} \dfrac{R(x)}{\omega - x}\,dx \end{cases}$

These two equations define the Hilbert transform.

We now decompose $f(t)$ into an even function $f_p(t)$ and an odd function $f_i(t)$:

$$\begin{cases} f_p(t) = \dfrac{f(t) + f(-t)}{2} \\ f_i(t) = \dfrac{f(t) - f(-t)}{2} \end{cases}$$

We want to demonstrate that in the case of a causal function, the following is true:

$$\begin{cases} f_p(t) = f_i(t)sign(t) \\ f_i(t) = f_p(t)sign(t) \end{cases}$$

The function defined by: $sign(t) = \begin{cases} +1 \text{ when } t > 0 \\ -1 \text{ when } t > 0 \end{cases}$

2.3. Solutions to the exercises

2.3.1. *Exercise 2.1*

The Fourier transform of $f(t)$ appears as follows:

$$F(j\omega)=\int_{-\infty}^{+\infty} f(t)e^{-j\omega t}dt$$

By replacing $e^{-j\omega t}$ with $\cos\omega t - j\sin\omega t$, we find the desired result.

If $f(t)$ is even, $f_{(t)}\cos\omega t$ is also even whereas $f(t)\sin\omega t$ is odd.

From this, we establish the following:

$$\int_{-\infty}^{+\infty} f(t)\sin\omega t dt = 0$$

$$\int_{-\infty}^{+\infty} f(t)\cos\omega t dt = 2\int_{0}^{+\infty} f(t)\cos\omega t dt$$

The Fourier transform then appears as follows:

$$\boxed{F(j\omega)=2\int_{0}^{+\infty} f(t)\cos\omega t dt}$$

This reasoning also applies to cases where $f(t)$ is an odd function.

We then obtain: $\boxed{F(j\omega)=-2j\int_{0}^{+\infty} f(t)\sin\omega t dt}$

Let us now consider the case of $f(t)$ being real and even. The following can be said:

$$F(j\omega)=2\int_{0}^{+\infty} f(t)\cos\omega t dt$$

is a function that is said to be real and even in ω.

We know that:

$$u(t) \xleftrightarrow{TF} \frac{1}{j\omega} + \pi\delta(\omega) \Rightarrow u(t-\tau) \xleftrightarrow{TF} \frac{e^{-j\omega\tau}}{j\omega} + \pi\delta(\omega)$$

According to the sampling properties of "function" δ, we also know that $e^{-j\omega\tau}\delta(\omega) = \delta(\omega)$.

Thus: $$\boxed{F_5(j\omega) = \frac{1}{j\omega}\left(1 + e^{-j\omega} + e^{-j2\omega} - 2e^{-j3\omega} - e^{-j4\omega}\right)}$$

6) $f_6(t) = \begin{cases} 1-t & t \in \left[0, \frac{3}{2}\right] \\ t-2 & t \in \left[\frac{3}{2}, 2\right] \\ 0 & t \in [2, \infty] \end{cases}$ with an even function.

We can use the results from exercise 2.1.

$$F_6(j\omega) = 2\int_0^{\tau} f_6(t)\cos\omega t\,dt$$

$$= 2\left[\int_0^{3/2} (1-t)\cos\omega t\,dt + \int_{3/2}^{2} (t-2)\cos\omega t\,dt\right]$$

After integrating part-by-part, we obtain the following result:

$$\boxed{F_6(j\omega) = \frac{1}{\omega^2}\left(1 - 2\cos\frac{3}{2}\omega + \cos 2\omega\right)}$$

7) $f_7(t)$ is expressed as:

$$f_7(t) = f_1(t+2) + f_2(t)$$

by tethering the value τ to *1* for $f_2(t)$ and to *2* for $f_1(t)$:

– For $\tau = 1 \Rightarrow f_2(t) \xleftrightarrow{TF} F_2(j\omega) = \frac{1}{\omega^2}\left(1 - j\omega - e^{-j\omega}\right)$

– For $\tau = 2 \Rightarrow f_1(t) \xleftrightarrow{TF} F_1(j\omega) = \frac{1}{2\omega^2}\left((1 + 2j\omega)e^{2j\omega} - 1\right)$

Thus: $F_7(j\omega) = F_1(j\omega)e^{-2j\omega}F_2(j\omega)$

Or rather: $\boxed{F_7(j\omega) = \frac{1}{2\omega^2}\left(3 - 2e^{-j\omega} - e^{-2j\omega}\right)}$

8) $f_8(t) = \begin{cases} \frac{t}{\tau} & t \in [-\tau, +\tau] \\ 0 & t \notin [-\tau, +\tau] \end{cases}$

This function is odd which means: $F_8(j\omega) = -2j\int_0^{\infty} f_8(t)\sin\omega t dt$, meaning that $F_8(j\omega) = -2j\int_0^{\tau} \frac{t}{\tau}\sin\omega t dt$

After part by part integration, we obtain:

$$\boxed{F_8(j\omega) = \frac{2j}{\tau\omega^2}\left(\omega\tau\cos\omega\tau - \sin\omega\tau\right)}$$

9) $f_9(t) = \begin{cases} \cos & t \in \left[0, \pi/2\right] \\ 0 & t \in \left[\pi/2, \infty\right[\end{cases}$

$f_9(t)$ is an even function, which is why: $F_9(j\omega) = 2\int_0^{\infty} f_9(t)\cos\omega t dt$.

This means that after integrating using the expression of the product of two cosines, the following is true:

$$F_9(j\omega)=\frac{\pi}{2}\left[\frac{\sin(\omega+1)\frac{\pi}{2}}{(\omega+1)\frac{\pi}{2}}+\frac{\sin(\omega-1)\frac{\pi}{2}}{(\omega-1)\frac{\pi}{2}}\right]$$

We can perform this calculation over again using the Dirac impulse method. This is only possible for functions $f_3(t)$, $f_4(t)$, $f_5(t)$, $f_6(t)$ and $f_7(t)$ which are the only ones that give a series of Dirac impulses through successive derivatives.

We will indicate here the methods for functions $f_3(t)$ and $f_5(t)$:

– For $f_3(t)$:

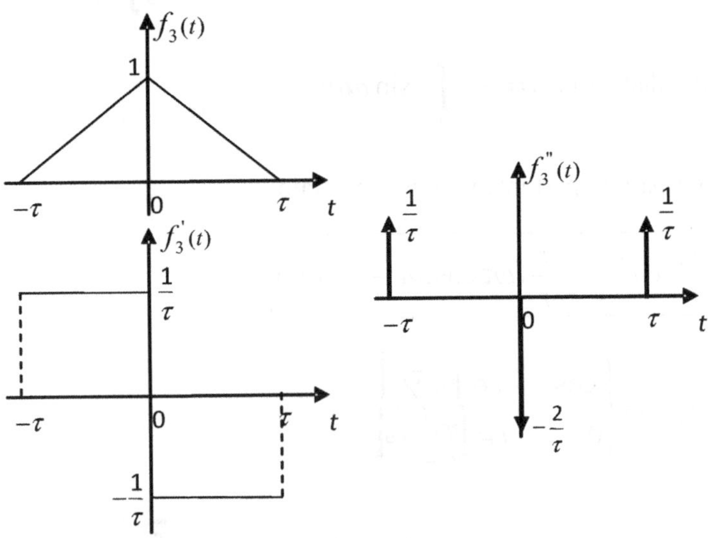

$$f_3''(t)=\frac{1}{\tau}\left(\delta(t+\tau)-2\delta(t)+\delta(t-\tau)\right)$$

Either using a Fourier transform:

$$-\omega^2 F_3(j\omega) = \frac{1}{\tau}\left(e^{j\omega\tau} - 2 + e^{-j\omega\tau}\right)$$

This gives us the following result:

$$\boxed{F_3(j\omega) = \frac{2}{\tau\omega^2}(1 - \cos\omega\tau)}$$

identical to the one we found using the direct calculation.

– For $f_5(t)$

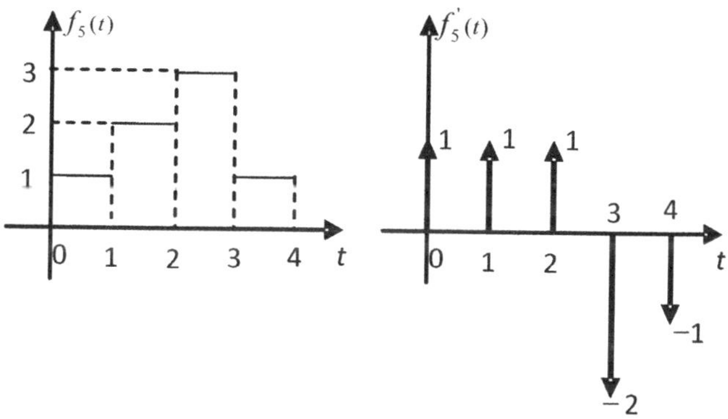

$$f_5'(t) = \delta(t) + \delta(t-1) + \delta(t-2) - 2\delta(t-3) + \delta(t-4)$$

which gives us $j\omega F_5(j\omega) = 1 + e^{-j\omega} + e^{-2j\omega} - 2e^{-3j\omega} - e^{-4j\omega}$.

Note that in this example we only derived once.

For this method to be applicable, it is important to check every time that:

$$F(j\omega)\Big|_{\omega=0} = \int_{-\infty}^{+\infty} f(t)dt = 0$$

In cases where $f(t)$ is a series of impulses, this is the same as checking that the sum of the weights of the impulses is equal to 0.

2.3.3. *Exercise 2.3*

The Fourier transform of $\delta(t)$ is $F(j\omega) = 1$.

Applying the definition for the inverse Fourier transform:

$$f(t) = \frac{1}{2\pi} \int_{-\infty}^{+\infty} F(j\omega) e^{j\omega t} d\omega$$

we obtain: $$\boxed{\delta(t) = \frac{1}{2\pi} \int_{-\infty}^{+\infty} e^{j\omega t} d\omega}$$

What's more: $e^{j\omega t} = \cos\omega t + j\sin\omega t$

Function $\sin\omega t$ is odd, meaning that: $\int_{-\infty}^{+\infty} \sin\omega t \, d\omega = 0$

From this we deduce that: $$\boxed{\delta(t) = \frac{1}{\pi} \int_{0}^{+\infty} \cos\omega t d\omega}$$

2.3.4. *Exercise 2.4*

Since the calculations of the different transforms present no difficulties, we will only present the results.

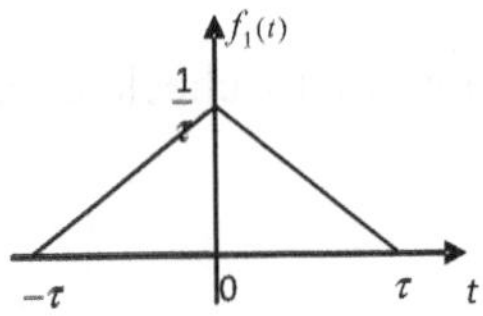

$$f_1(t) = \begin{cases} \frac{1}{\tau}\left(1 - \frac{t}{\tau}\right) & |t| < \tau \\ 0 & |t| > \tau \end{cases} \xleftrightarrow{TF} F_1(j\omega) = \left(\frac{\sin\frac{\omega\tau}{2}}{\frac{\omega\tau}{2}} \right)^2$$

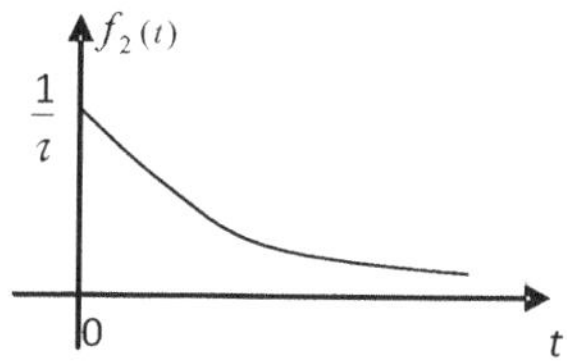

$$f_2(t) = \frac{1}{\tau} e^{-\frac{t}{\tau}} u(t) \xleftrightarrow{TF} F_2(j\omega) = \frac{1}{1 + j\omega\tau}$$

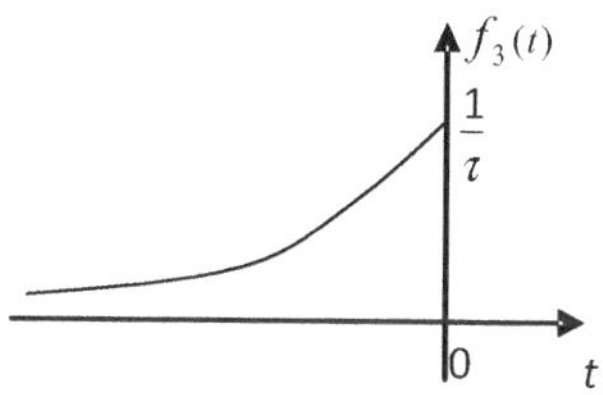

$$f_3(t) = \frac{1}{\tau} e^{\frac{t}{\tau}} u(-t) \xleftrightarrow{TF} F_3(j\omega) = \frac{1}{1 - j\omega\tau}$$

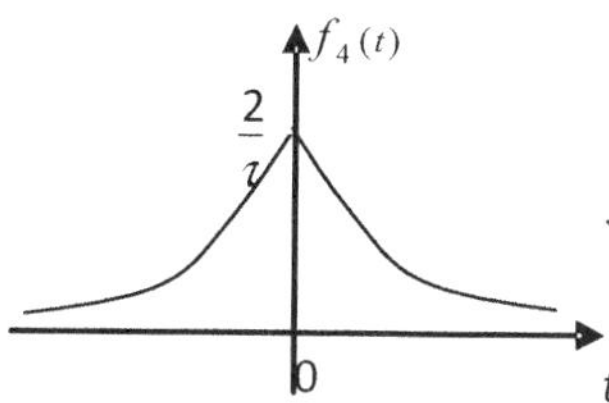

$$f_4(t) = \frac{1}{2\tau} e^{-\frac{|t|}{\tau}} \xleftrightarrow{TF} F_4(j\omega) = \frac{1}{1 + (\omega\tau)^2}$$

When $\tau \to 0$, the Fourier transforms become equal to 1.

$$F_1(j\omega) = F_1(j\omega) = F_1(j\omega) = F_4(j\omega) = 1$$

When $\tau \to 0$, the functions see their width become null and their amplitudes become infinite while maintaining an area included under the curve constant and equal to 1. These functions approach the Dirac impulses.

$$\lim_{\tau \to 0} f_1(t) = \lim_{\tau \to 0} f_2(t) = \lim_{\tau \to 0} f_3(t) = \lim_{\tau \to 0} f_4(t) = \delta(t)$$

It can therefore be said that the Fourier transforms approach the transform of the Dirac impulses.

2.3.5. *Exercise 2.5*

Consider $f(t) = \frac{\alpha}{\pi}\left(\frac{\sin \alpha t}{\alpha t}\right)^2$. We will plot the shape of the curve and calculate the area under the curve.

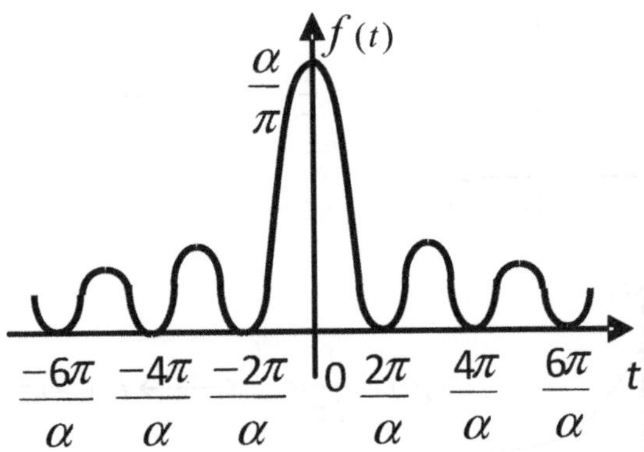

$$Air = \int_{-\infty}^{+\infty} \frac{\alpha}{\pi}\left(\frac{\sin \alpha t}{\alpha t}\right)^2 dt = \frac{\alpha}{\pi}\int_{-\infty}^{+\infty}\left(\frac{\sin x}{x}\right)^2 \frac{dx}{\alpha} = 1$$

When $\alpha \to \infty$, the width of the curve approaches 0, the amplitude approaches infinity and the surface remains constant and equal to 1. The function approaches the Dirac impulse $\delta(t)$.

This gives us: $$\boxed{\delta(t) = \lim_{\alpha \to \infty} \frac{\alpha}{\pi}\left(\frac{\sin \alpha t}{\alpha t}\right)^2}$$

2.3.6. *Exercise 2.6*

We calculate $f(t) * u(t) = \int_{-\infty}^{+\infty} f(\tau)u(t-\tau)d\tau$.

Using the fact that: $u(t-\tau)=\begin{cases}0 \text{ when } & t>\tau\\ 1 \text{ when } & t<\tau\end{cases}$

we obtain: $$f(t)*u(t)=\int_{-\infty}^{t} f(\tau)u(t-\tau)d\tau=\int_{-\infty}^{t} f(\tau)d\tau$$

The Fourier transform of the convolution is:

$$f(t)*u(t)\xleftrightarrow{TF} F(j\omega)\left(\frac{1}{j\omega}+\pi\delta(\omega)\right)=\frac{F(j\omega)}{j\omega}+\pi F(j\omega)\big|_{\omega=0}\delta(\omega)$$

Thus: $$\boxed{\int_{-\infty}^{t} f(\tau)d\tau \xleftrightarrow{TF} \frac{F(j\omega)}{j\omega}+\pi F(j\omega)\big|_{\omega=0}\delta(\omega)}$$

If $F(j\omega)\big|_{\omega=0}=0$, we end up with the result indicated in the theoretical background section.

2.3.7. *Exercise 2.7*

We represent the shape of these functions:

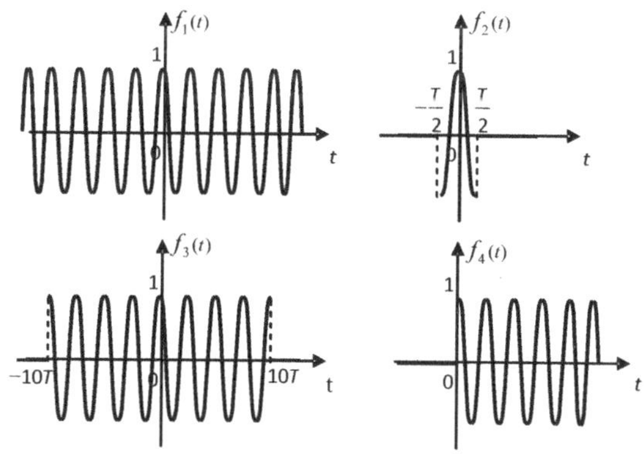

We easily obtain the following results:

$$F_1(j\omega) = \pi\left(\delta(\omega-\omega_0)+\delta(\omega+\omega_0)\right)$$

$$F_2(j\omega) = \frac{\pi}{\omega_0}\left[\frac{\sin\pi\left(\dfrac{\omega}{\omega_0}-1\right)}{\pi\left(\dfrac{\omega}{\omega_0}-1\right)}+\frac{\sin\pi\left(\dfrac{\omega}{\omega_0}+1\right)}{\pi\left(\dfrac{\omega}{\omega_0}+1\right)}\right]$$

$$F_3(j\omega) = \frac{20\pi}{\omega_0}\left[\frac{\sin 20\pi\left(\dfrac{\omega}{\omega_0}-1\right)}{20\pi\left(\dfrac{\omega}{\omega_0}-1\right)}+\frac{\sin 20\pi\left(\dfrac{\omega}{\omega_0}+1\right)}{20\pi\left(\dfrac{\omega}{\omega_0}+1\right)}\right]$$

$$F_4(j\omega) = \frac{\pi}{2}\left(\delta(\omega-\omega_0)+\delta(\omega+\omega_0)\right) - j\frac{\omega}{\omega^2-\omega_0^2}$$

We represent these different spectra:

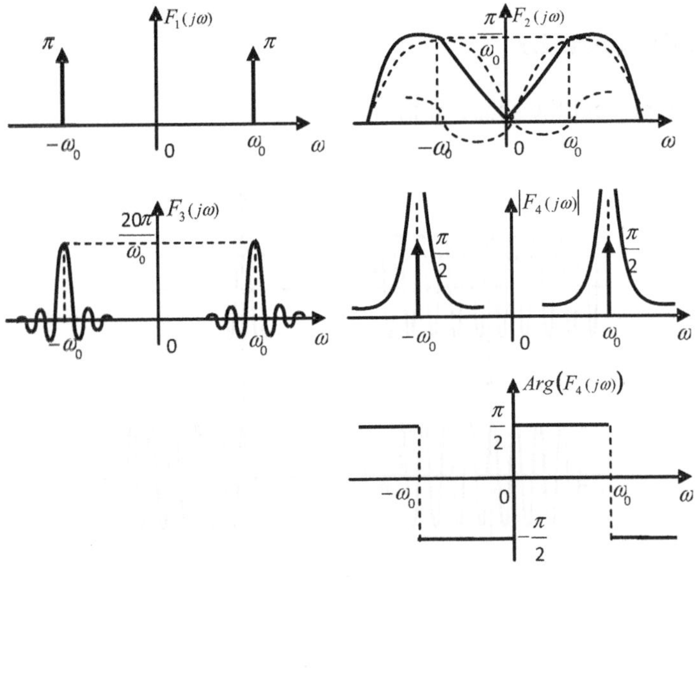

The Fourier transforms of $f_2(t)$ and $f_3(t)$ highlight an increasing concentration of energy around pulsations $\pm\omega_0$ as the number of periods taken into account increases.

This amounts to using functions that increasingly approach $f_1(t)$. As they approach the limit, functions $\dfrac{\sin x}{x}$ approach impulse $\delta(x)$ and the spectrum becomes a line spectrum.

Function $f_4(t)$ is causal and thus is null if $t<0$, its spectrum contains a continuous part outside of $\pm\omega_0$ so as to cancel out this function on the interval $]-\infty,0]$

2.3.8. *Exercise 2.8*

The first case will be treated using the duality property and the second using direct calculation:

– We already know the following result:

$$f(t) = Cr_\tau(t) \overset{TF}{\longleftrightarrow} F(j\omega) = \tau \frac{\sin \frac{\omega\tau}{2}}{\frac{\omega\tau}{2}}$$

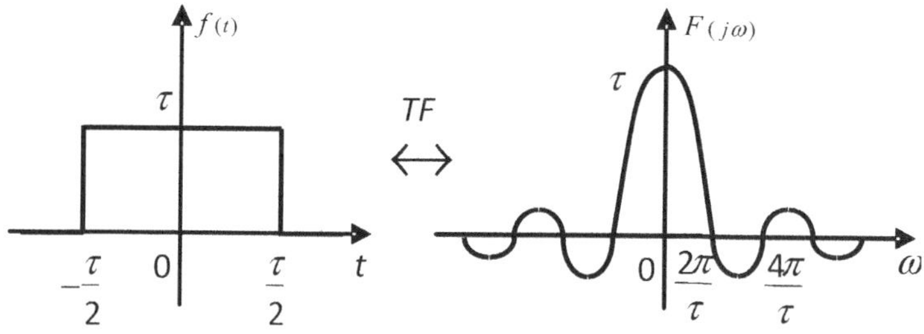

Using the duality property:

$$f(t) \xleftrightarrow{TF} F(j\omega) \Rightarrow F(t) \xleftrightarrow{TF} 2\pi f(-j\omega)$$

we obtain: $F(t) = \omega_0 \dfrac{\sin\dfrac{\omega_0 t}{2}}{\dfrac{\omega_0 t}{2}} \xleftrightarrow{TF} 2\pi f(-j\omega) = 2\pi Cr_{\omega_0}(\omega)$

The reverse transform $Cr_{2\omega_0}(\omega)$ appears:

$$\boxed{Cr_{2\omega_0}(\omega) \xleftrightarrow{TF} \frac{\omega_0}{\pi}\frac{\sin\omega_0 t}{\omega_0 t}}$$

If we now introduce argument $\varphi_{(\omega)} = -\omega t_0$, using the time-domain translation translation property we obtain:

$$\boxed{Cr_{2\omega_0}(\omega)e^{-j\omega t_0} \xleftrightarrow{TF} \frac{\omega_0}{\pi}\frac{\sin\omega_0(t-t_0)}{\omega_0(t-t_0)}}$$

This function appears as the following graph.

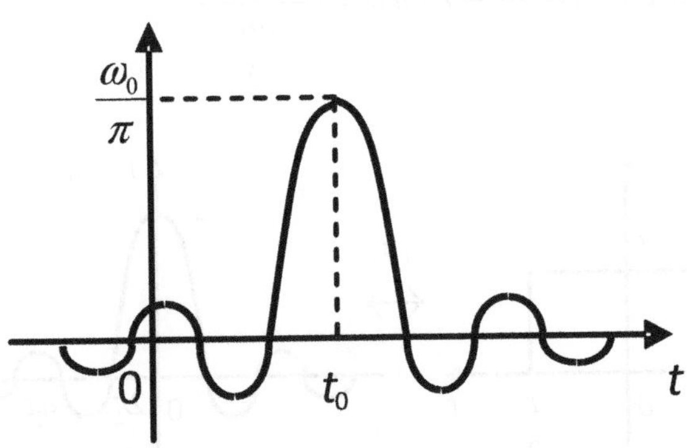

– We express $F_2(j\omega)$ in an analytical form:

$$F_2(j\omega)=\begin{cases}0 & \omega\in\left]-\infty,-\omega_0\right]\\ j & \omega\in\left[-\omega_0,0\right]\\ -j & \omega\in\left[0,+\omega_0\right]\\ 0 & \omega\in\left[+\omega_0,+\infty\right[\end{cases}$$

We use equation: $f(t)=\dfrac{1}{2\pi}\displaystyle\int_{-\infty}^{+\infty}F(j\omega)e^{j\omega t}d\omega$

Which gives us $f_2(t)=\dfrac{1}{2\pi}\left(\displaystyle\int_{-\omega_0}^{0}je^{j\omega t}d\omega+\int_{0}^{\omega_0}-je^{j\omega t}d\omega\right)$

Or: $\boxed{f_2(t)=\dfrac{1}{\pi}\dfrac{1-\cos\omega_0 t}{t}}$

The graph is as follows.

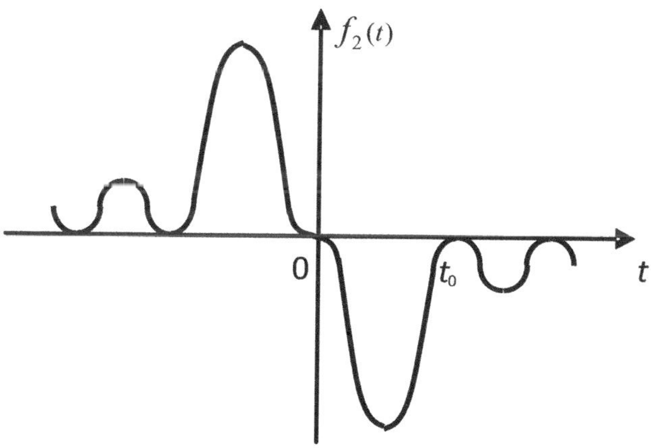

2.3.9. Exercise 2.9

– Function $f_1(t)$ gives $f_1(t) = Cr_{\frac{2\pi}{5}}(t)\cos 20t$:

We know that: $Cr_{\frac{2\pi}{5}}(t) \xleftrightarrow{TF} \dfrac{2\pi}{5}\dfrac{\sin\frac{2\pi}{5}\omega}{\frac{2\pi}{5}\omega}$

Using the theorem of frequency-domain translation:

$$F_1(j\omega) = \frac{\pi}{5}\left(\frac{\sin\frac{2\pi}{5}(\omega-20)}{\frac{2\pi}{5}(\omega-20)} + \frac{\sin\frac{2\pi}{5}(\omega+20)}{\frac{2\pi}{5}(\omega+20)}\right)$$

– Function $f_2(t)$ can be written in the following form: $f_2(t) = g(t)\cos 20t$

Function $g(t)$ being: $g(t) = \begin{cases} \dfrac{40t}{9\pi} & t \in \left]-\dfrac{9\pi}{40}, +\dfrac{9\pi}{40}\right] \\ 0 & t \notin \left]-\dfrac{9\pi}{40}, +\dfrac{9\pi}{40}\right] \end{cases}$

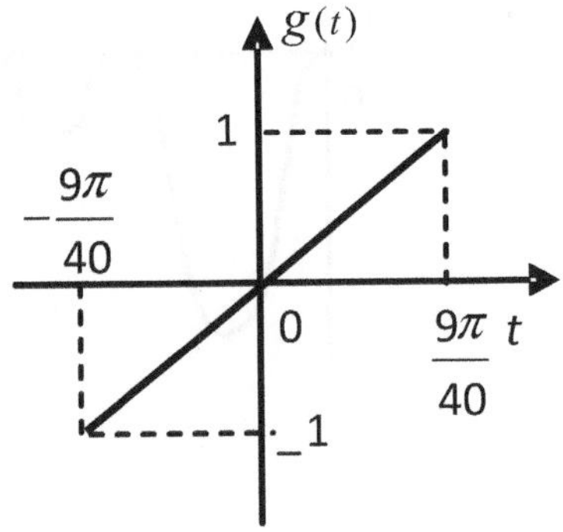

The transform of $g(t)$ was calculated already in exercise 2.2 (function $f_8(t)$).

$$g(t) \xleftrightarrow{TF} G(j\omega) = \frac{2j}{\frac{9\pi}{40}\omega^2}\left(\frac{9\pi}{40}\omega\cos\frac{9\pi}{40}\omega - \sin\frac{9\pi}{40}\omega\right)$$

$$= 2j\frac{1}{\omega}\cos\frac{9\pi}{40}\omega - j\frac{80}{9\pi}\frac{1}{\omega^2}\sin\frac{9\pi}{40}\omega$$

By applying the property of frequency-domain translation, we obtain:

$$F_2(j\omega) = j\left(\frac{\cos\frac{9\pi}{40}(\omega-20)}{(\omega-20)} + \frac{\cos\frac{9\pi}{40}(\omega+20)}{(\omega+20)}\right) - j\frac{40}{9\pi}\left(\frac{\sin\frac{9\pi}{40}(\omega-20)}{(\omega-20)^2} + \frac{\sin\frac{9\pi}{40}(\omega+20)}{(\omega+20)^2}\right)$$

– Function $f_3(t)$ can be written in the following form:

$$f_3(t) = g(t)\cos 200\pi t$$

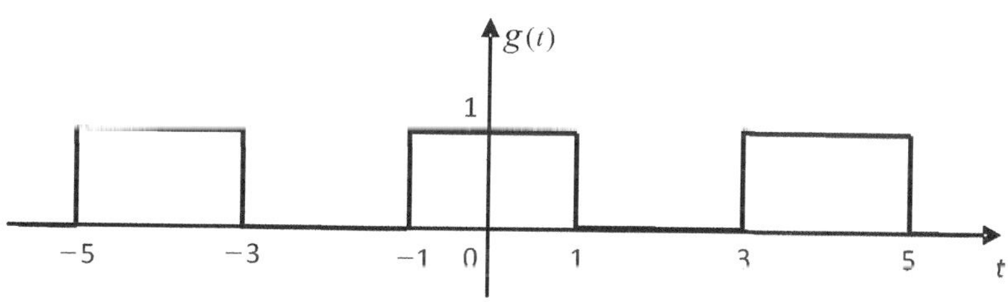

$g(t)$ is a square periodic function and we know its Fourier decomposition:

$$g(t) = \sum_{n=-\infty}^{n=+\infty} C_n e^{jn\omega_0 t} \quad \text{with} \quad C_n = \frac{\tau}{T}\frac{\sin n\frac{\omega_0\tau}{2}}{n\frac{\omega_0\tau}{2}}$$

In this case $\tau = 2 \quad T = 4 \quad \Rightarrow \omega_0 = \frac{\pi}{2}$, hence: $C_n = \frac{1}{2}\frac{\sin n\frac{\pi}{2}}{n\frac{\pi}{2}}$

From this we establish the Fourier transform of $g(t)$:

$$G(j\omega) = 2\pi \sum_{n=-\infty}^{n=+\infty} C_n \delta(\omega - n\omega_0)$$

$$G(j\omega) = \pi \sum_{n=-\infty}^{n=+\infty} \frac{\sin n\frac{\pi}{2}}{n\frac{\pi}{2}} \delta(\omega - n\frac{\pi}{2})$$

By applying the frequency-domain translation property, we obtain:

$$\boxed{F_3(j\omega) = \frac{\pi}{2} \sum_{n=-\infty}^{n=+\infty} \frac{\sin n\frac{\pi}{2}}{n\frac{\pi}{2}} \left(\delta(\omega - 200\pi - n\frac{\pi}{2}) + \delta(\omega + 200\pi - n\frac{\pi}{2}) \right)}$$

– Function $f_4(t)$ can be written in the following form:

$f_4(t) = g(t)\cos 200\pi t$

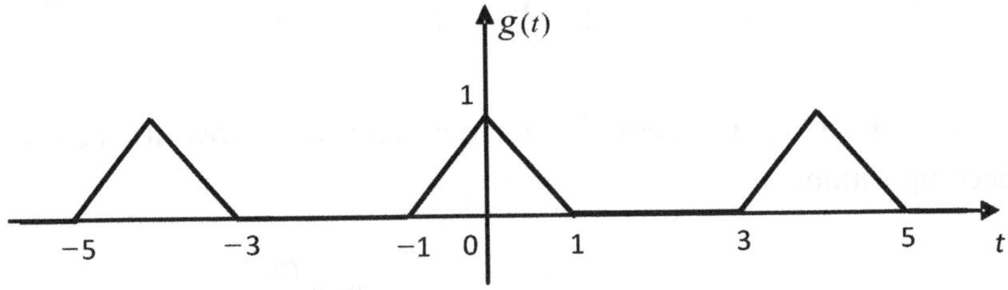

where $g(t)$ is a saw-wave type periodic function.

The process is the same as for the previous case, so we will just indicate the results below:

$$G(j\omega) = \frac{\pi}{2} \sum_{n=-\infty}^{n=+\infty} \left(\frac{\sin n\frac{\pi}{4}}{n\frac{\pi}{4}} \right)^2 \delta(\omega - n\frac{\pi}{2})$$

$$\boxed{F_4(j\omega) = \frac{\pi}{4} \sum_{n=-\infty}^{n=+\infty} \left(\frac{\sin n\frac{\pi}{4}}{n\frac{\pi}{4}} \right)^2 \left(\delta(\omega - 200\pi - n\frac{\pi}{2}) + \delta(\omega + 200\pi - n\frac{\pi}{2}) \right)}$$

2.3.10. *Exercise 2.10*

– Consider the following function $G(j\omega)$:

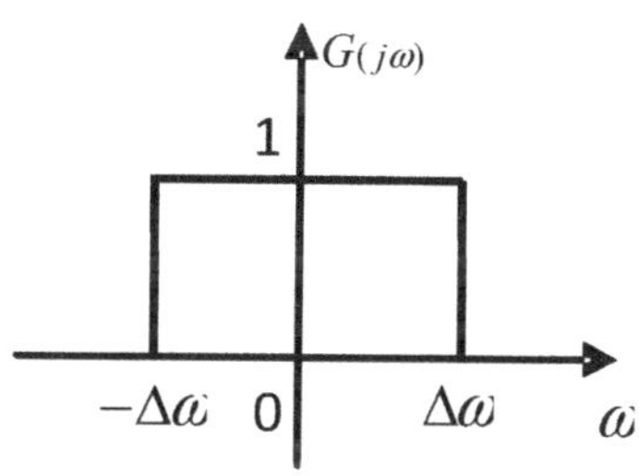

Consider $g(t)$ the inverse Fourier transform of $G(j\omega)$

We can then write:

$$g(t)\cos\omega_0 t \xleftrightarrow{TF} F_1(j\omega) = \frac{1}{2}\left(G(\omega - \omega_0) + G(\omega + \omega_0)\right)$$

This equation is valid because $2\Delta\omega < \omega_0$, which avoids overlap.

We know $g(t)$ (see Table 2.1): $g(t) = \frac{\Delta\omega}{\pi}\frac{\sin\Delta\omega t}{\Delta\omega t}$

Therefore: $$F_1(j\omega) \xleftrightarrow{TF} f_1(t) = \frac{\Delta\omega}{\pi}\frac{\sin\Delta\omega t}{\Delta\omega t}\cos\omega_0 t$$

– The second case can be approached in exactly the same way:

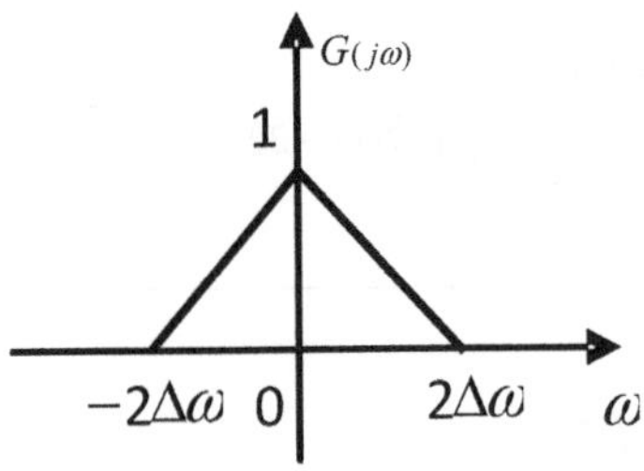

$$g(t) = \frac{\Delta\omega}{\pi}\left(\frac{\sin\Delta\omega t}{\Delta\omega t}\right)^2$$

$$F_2(j\omega) \xleftrightarrow{TF} f_2(t) = \frac{\Delta\omega}{\pi}\left(\frac{\sin\Delta\omega t}{\Delta\omega t}\right)^2 \cos\omega_0 t$$

2.3.11 *Exercise 2.11*

Let us determine the Fourier transform of $f(t+T)$:

$$f(t+T) \xleftrightarrow{TF} F_1(j\omega) = \int_{-\infty}^{+\infty} f(t+T)e^{-j\omega t}dt$$

We pose $x = t + T$, which gives us:

$$F_1(j\omega) = \int_{-\infty}^{+\infty} f(x)e^{-j\omega(x-T)}dx = e^{j\omega T}\int_{-\infty}^{+\infty} f(x)e^{-j\omega x}dx = e^{j\omega T}F(j\omega)$$

Therefore $f(t+T) \overset{TF}{\leftrightarrow} e^{j\omega T} F(j\omega)$. This is the result we obtain when applying the time-domain translation.

In the same way, we obtain: $f(t-T) \overset{TF}{\leftrightarrow} e^{-j\omega T} F(j\omega)$

Or by combining both relations:

$$\boxed{\begin{array}{l} f(t+T)+f(t-T) \overset{TF}{\longleftrightarrow} 2\cos\omega T \; F(j\omega) \\ f(t+T)-f(t-T) \overset{TF}{\longleftrightarrow} 2j\sin\omega T \; F(j\omega) \end{array}}$$

– Function $f_1(t)$ is written:

$$f_1(t) = g(t-2) + g(t+2)$$

$g(t)$ being a square function: $g(t) = Cr_1(t)$

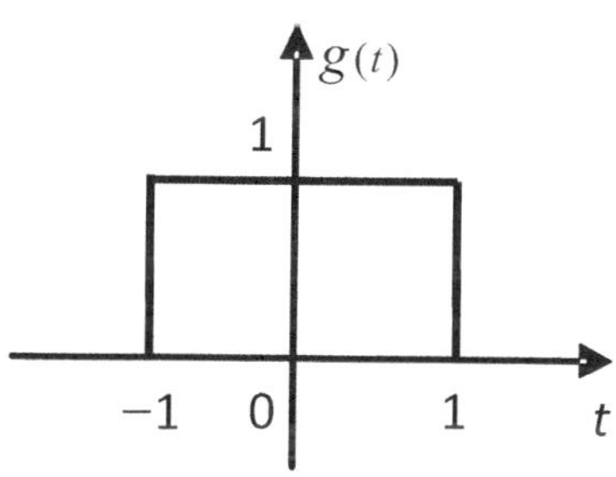

$$g(t) \overset{TF}{\longleftrightarrow} G(j\omega) = 2\frac{\sin\omega}{\omega}$$

Therefore: $\boxed{F_1(j\omega) = 4\cos 2\omega \frac{\sin\omega}{\omega}}$

– Function $f_2(t)$ appears as follows:

$$f_2(t) = g(t-\frac{3}{2}) + g(t+\frac{3}{2})$$

$g(t)$ being the following function:

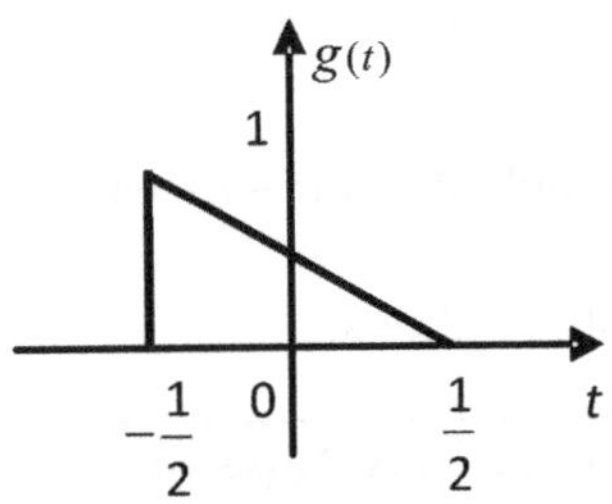

In exercise 2.1, we calculated the Fourier transform of. $g_1(t) = g(t - \frac{1}{2})$:

$$g_1(t) \xleftrightarrow{TF} G_1(j\omega) = \frac{1}{\omega^2}\left(1 - j\omega - e^{-j\omega}\right)$$

The Fourier transform of $g(t)$ is therefore:

$$G_1(j\omega) = G_1(j\omega)e^{j\frac{\omega}{2}} = \frac{2j}{\omega^2}\left(\sin\frac{\omega}{2} - \frac{\omega}{2}e^{j\frac{\omega}{2}}\right)$$

This gives us the Fourier transform of $f_2(t)$:

$$\boxed{F_2(j\omega) = \frac{4j}{\omega^2}\cos\frac{3\omega}{2}\left(\sin\frac{\omega}{2} - \frac{\omega}{2}e^{j\frac{\omega}{2}}\right)}$$

2.3.12 *Exercise 2.12*

– Direct calculation:

$f(t)$ is odd so we can write:

$$F(j\omega) = -2j\int_0^{\tau} f(t)\sin\omega t\, dt = -2j\int_0^{\tau}(-1)\sin\omega t\, dt$$

or: $$\boxed{F(j\omega) = 2j\tau\left(\frac{1-\cos\omega\tau}{\omega\tau}\right)}$$

– Deriving $f(t)$, we obtain:

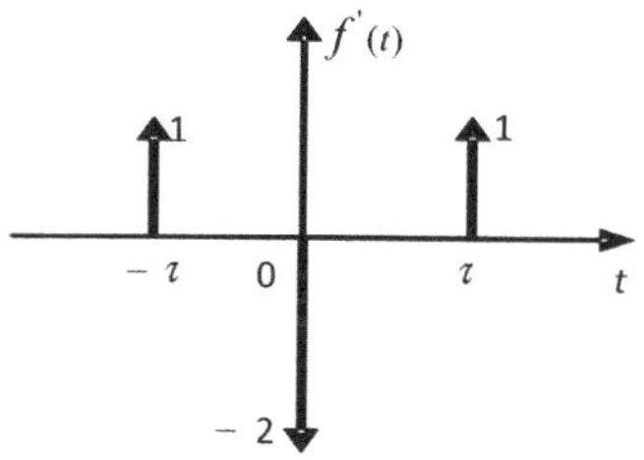

$$f'(t) = \delta(t+\tau) - 2\delta(t) + \delta(t-\tau)$$

$$f'(t) \xleftrightarrow{TF} F_1(j\omega) = e^{j\omega\tau} - 2 + e^{-j\omega\tau} = 2(\cos\omega\tau - 1)$$

As $f(t) = \int_{-\infty}^{t} f'(\tau)d\tau$, we can establish the following as true:

$$f(t) \xleftrightarrow{TF} F(j\omega) = \frac{F_1(j\omega)}{j\omega} \text{ car } F_1(j\omega)\big|_{\omega=0} = 0$$

Which means that: $$\boxed{F(j\omega) = 2j\tau\left(\frac{1-\cos\omega\tau}{\omega\tau}\right)}$$

– If we integrate $f(t)$, we obtain:

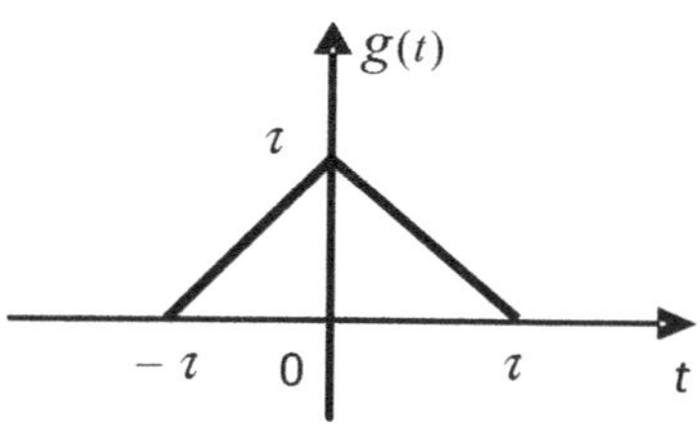

$$g(t) = \int_{-\infty}^{t} f(\tau) d\tau$$

The Fourier transform $G(j\omega)$ of $g(t)$ was calculated in exercise 2.2:

$$G(j\omega) = \tau \left(\frac{\sin \frac{\omega\tau}{2}}{\frac{\omega\tau}{2}} \right)^2$$

$$f(t) = \frac{dg(t)}{dt} \Rightarrow F(j\omega) = j\omega G(j\omega)$$

By arranging this: $\boxed{F(j\omega) = 2j\tau \left(\frac{1 - \cos \omega\tau}{\omega\tau} \right)}$

– Considering a square function $Cr_\tau(t)$, we can set:

$$f(t) = Cr_\tau(t + \frac{T}{2}) - Cr_\tau(t - \frac{T}{2})$$

$$Cr_\tau(t) \xleftrightarrow{TF} \tau \frac{\sin \frac{\omega\tau}{2}}{\frac{\omega\tau}{2}}$$

Therefore $F(j\omega) = \tau \dfrac{\sin \frac{\omega\tau}{2}}{\frac{\omega\tau}{2}} \left(e^{j\frac{\omega\tau}{2}} - e^{j\frac{\omega\tau}{2}} \right) = 2j\tau \dfrac{\left(\sin \frac{\omega\tau}{2} \right)^2}{\omega\tau}$

If we rearrange this: $\boxed{F(j\omega) = 2j\tau \frac{1 - \cos \omega\tau}{\omega\tau}}$

2.3.13 *Exercise 2.13*

"Function" $\delta_T(t)$ is periodic and therefore admits an exponential Fourier decomposition.

$$C_n = \frac{1}{T}\int_{-\frac{T}{2}}^{+\frac{T}{2}} \delta(t)e^{-jn\omega_0 t}dt = \frac{1}{T}\left[e^{-jn\omega_0 t}\right]_{t=0} = \frac{1}{T} = \frac{\omega_0}{2\pi}$$

The Fourier transform the appears as follows:

$$F(\omega) = 2\pi\sum_{n=-\infty}^{n=+\infty} C_n\delta(\omega - n\omega_0)$$

which gives: $$\boxed{F(\omega) = \omega_0\sum_{n=-\infty}^{n=+\infty} \delta(\omega - n\omega_0)}$$

This can then be written under the following form:

$$\boxed{\delta_T(t) \xleftrightarrow{TF} \delta_{\omega_0}(\omega)}$$

2.3.14 *Exercise 2.14*

Let us use the time-domain translation property:

$$f(t - pT) \xleftrightarrow{TF} F(j\omega)e^{jpT}$$

By calling $F_n(j\omega)$ the Fourier transform of $f_n(t)$

$$F_n(j\omega) = \sum_{p=-n}^{p=+n} F(j\omega)e^{-jpT\omega}$$

Using the mathematical relation of the exercise, we obtain:

$$F_n(j\omega) = F(j\omega)\frac{\sin(2n+1)\frac{\omega T}{2}}{\frac{\omega T}{2}}$$

When $n \to \infty \Rightarrow F_n(j\omega)$ becomes $F(j\omega)\sum_{p=-\infty}^{p=+\infty} e^{-jpT\omega}$

But $\sum_{p=-\infty}^{p=+\infty} e^{-jpT\omega}$ is the Fourier transform of $\delta_T(t)$

Which can be written in the following form (see exercise 1.13):

$$\delta_T(t) \xleftrightarrow{TF} \omega_0 \sum_{p=-\infty}^{p=+\infty} \delta(\omega - p\omega_0)$$

Therefore: $$\lim_{n\to\infty} f_n(t) \xleftrightarrow{TF} \omega_0 \sum_{p=-\infty}^{p=+\infty} F(p\omega_0)\delta(\omega - p\omega_0)$$

When $n \to \infty$, $f_n(t)$ approaches a periodic "function", and we find the Fourier transform of a periodic function.

2.3.15 *Exercice 2.15*

– We calculate the Fourier transform using the Dirac impulse method:

$$f''(t) = \frac{A}{\tau}\left(\delta(t-(t_0-\frac{T}{2}-\frac{\tau}{2})) - \delta(t-(t_0-\frac{T}{2}+\frac{\tau}{2})) - \delta(t-(t_0+\frac{T}{2}-\frac{\tau}{2})) + \delta(t-(t_0+\frac{T}{2}+\frac{\tau}{2}))\right)$$

Using the Fourier transform of both sides:

$$(-j\omega)^2 F(j\omega) = \frac{A}{\tau}e^{-j\omega t_0}\left(e^{j\frac{\omega}{2}(T+\tau)} - e^{j\frac{\omega}{2}(T-\tau)} - e^{-j\frac{\omega}{2}(T-\tau)} + e^{-j\frac{\omega}{2}(T+\tau)}\right)$$

When we arrange this, we obtain:

$$\boxed{F(j\omega) = \frac{4A}{\tau\omega^2} e^{-j\omega t_0} \sin\frac{\omega T}{2} \sin\frac{\omega\tau}{2}}$$

Thus: $$\boxed{\begin{cases} |F(j\omega)| = ATe^{-j\omega t_0} \dfrac{\sin\frac{\omega T}{2}}{\frac{\omega T}{2}} \dfrac{\sin\frac{\omega\tau}{2}}{\frac{\omega\tau}{2}} \\ Arg\big(F(j\omega)\big) = -\omega t_0 \end{cases}}$$

– We calculate the different parameters.

$$C = \int_{-\infty}^{+\infty} |f(t)|dt = AT$$

$$V = \int_{-\infty}^{+\infty} \left|\frac{df(t)}{dt}\right| dt = 2A$$

$$O = \int_{-\infty}^{+\infty} \left|\frac{d^2 f(t)}{dt^2}\right| dt = \frac{4A}{\tau}$$

– We can write that:

$$\begin{cases} |F(j\omega)| \leq AT \\ |F(j\omega)| \leq \dfrac{2A}{|\omega|} \\ |F(j\omega)| \leq \dfrac{4A}{\tau} \end{cases}$$

We can trace the upper limits of $|F(j\omega)|$.

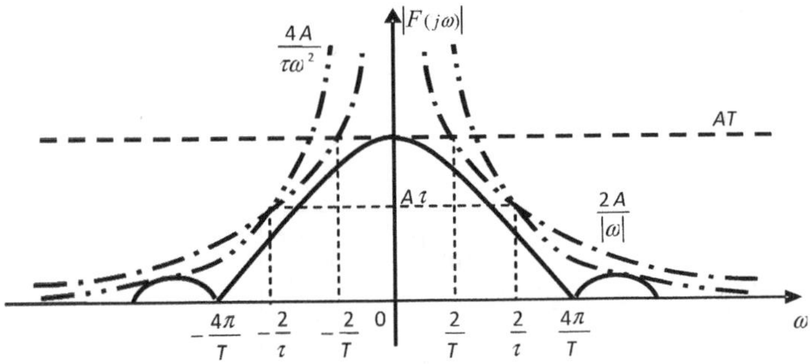

2.3.16 *Exercise 2.16*

Using the derivation in the frequency-domain property, we can state the following:

$$t^n f(t) \xleftrightarrow{TF} \frac{1}{(-j)^n} \frac{d^n F(j\omega)}{d\omega^n} = G(j\omega)$$

The moment of order n appears as:

$$m_n = \int_{-\infty}^{+\infty} t^n f(t) dt = G(j\omega)\Big|_{\omega=0}$$

Thus: $\boxed{m_n = j^n \left.\frac{d^n F(j\omega)}{d\omega^n}\right|_{\omega=0}}$

The Taylor decomposition of $F(j\omega)$ is written:

$$F(j\omega) = \sum_{n=0}^{n=+\infty} \left.\frac{d^n F(j\omega)}{d\omega^n}\right|_{\omega=0} \frac{\omega^n}{n!}$$

Hence this relation: $F(j\omega) = \sum_{n=0}^{n=+\infty} (-j)^n m_n \frac{\omega^n}{n!}$

2.3.17 *Exercise 2.17*

As function $f(t)$ is causal, functions $f_p(t)$ and $f_i(t)$ are not independent:

$$f_p(t)=\frac{f(t)+f(-t)}{2} \text{ and } f_i(t)=\frac{f(t)-f(-t)}{2}$$

If $t>0$ $\quad f(-t)=0 \qquad \Rightarrow f_p(t)=f_i(t)$

If $t<0$ $\quad f(t)=0 \qquad \Rightarrow f_p(t)=-f_i(t)$

This can be written as: $\boxed{f_p(t)=f_i(t)sign(t)}$

$f_p(t)$ and $f_i(t)$ are real so we can write:

$$f_p(t)\xleftrightarrow{TF}R(\omega) \qquad \text{real}$$
$$f_i(t)\xleftrightarrow{TF}jX(\omega) \qquad \text{imaginary}$$

Knowing that $sign(t)\xleftrightarrow{TF}\frac{2}{j\omega}$, we find $f_i(t)sign(t)\xleftrightarrow{TF}\frac{1}{2\pi}jX(\omega)*\frac{2}{j\omega}$

Taking into account $f_p(t)=f_i(t)sign(t)$, we can state the following, by using the Fourier transform of both members:

$$R(\omega)=\frac{1}{2\pi}\left(jX(\omega)*\frac{2}{j\omega}\right)=\frac{1}{2\pi}\int_{-\infty}^{+\infty}jX(x)\frac{2}{j(\omega-x)}dx$$

Or: $$\boxed{R(\omega) = \frac{1}{\pi} \int_{-\infty}^{+\infty} \frac{X(x)}{\omega - x} dx}$$

Using $f_i(t) = f_p(t) sign(t)$, we obtain:

$$\boxed{X(\omega) = -\frac{1}{\pi} \int_{-\infty}^{+\infty} \frac{R(x)}{\omega - x} dx}$$

3

Laplace Transform

3.1. Theoretical background

While Fourier transforms apply to functions included on an interval ranging between $]-\infty,+\infty[$, we are now going to look at causal functions that are null if $t<0$. Generalizing the Fourier transform to complex pulses $p=\sigma+j\omega$ and applying it to causal functions gives us the Laplace transform.

3.1.1. *Definition*

Let us consider a causal function $f(t)$ and a complex pulse. We define the unilateral Laplace transform using equations:

$$\begin{cases} F(p)=\int_0^{\infty} f(t)e^{-pt}dt \\ f(t)=\dfrac{1}{j2\pi}\int_{\sigma-j\infty}^{\sigma+j\infty} F(p)e^{pt}dt \end{cases} \qquad [3.1]$$

The symbolic notation is as follows:

$$\mathcal{L}\{f(t)\}=F(p) \quad \text{and} \quad f(t)=\mathcal{L}^{1}\{F(p)\}$$

$f(t)$ is a continuous sum of complex exponentials e^{pt}. The complex amplitude relative to these components is $F(p)$. The integral is calculated along the integration path $\sigma + j\omega$, ω ranging between $-\infty$ and $+\infty$. This rigorous calculation is performed using the theory of analytic functions and the residue theorem in complex analysis.

3.1.2. *Existence of the Laplace transform*

Let $f(t)$ be an exponential order function, meaning there exists a real positive number α such that:

$$\lim_{t \to \infty} \left(e^{-\alpha t} f(t) \right) = 0$$

In this case, $F(p)$ is defined for $\boxed{\Re e(p) = \sigma > \alpha}$ [3.2]

The value of α that ensures convergence is called the region of convergence (ROC).

3.1.2.1. *Switching from Laplace transform to Fourier transform*

It is possible to go from a Laplace transform to a Fourier transform by switching p to $j\omega$. However this result is no longer valid if $F(p)$ has poles on the imaginary axis. The Laplace transform generalizes the Fourier transform

3.1.3. *Properties of the Laplace transform*

3.1.3.1. *Linearity*

Consider $F_1(p) = \mathcal{L}\{f_1(t)\}$ and $F_{2(p)} = \mathcal{L}\{f_2(t)\}$, if λ_1 and λ_2 are constants:

$$\boxed{\lambda_1 F_1(p) + \lambda_2 F_2(p) = \mathcal{L}\{\lambda_1 f_1(t) + \lambda_2 f_2(t)\}} \quad [3.3]$$

3.1.3.2. *Scale factor*

If $F(p)=\mathcal{L}\{f(t)\}$ and a is real and positive:

$$\boxed{\frac{1}{a}F(\frac{p}{a})=\mathcal{L}\{f(at)\}} \quad [3.4]$$

3.1.3.3. *Translation in the time-domain*

If $F(p)=\mathcal{L}\{f(t)\}$ and τ is real and positive:

$$\boxed{F(p)e^{-p\tau}=\mathcal{L}\{f(t-\tau)\}} \quad [3.5]$$

3.1.3.4. *Translation in the frequency domain*

If $F(p)=\mathcal{L}\{f(t)\}$ and p_0 is a given complex pulse:

$$\boxed{F(p-p_0)=\mathcal{L}\{f(t)e^{p_0 t}\}} \quad [3.6]$$

3.1.3.5. *Derivation in the frequency-domain*

If $F(p)=\mathcal{L}\{f(t)\}$:

$$\boxed{\frac{d^n F(p)}{dp^n}=\mathcal{L}\{(-t)^n f(t)\}} \quad [3.7]$$

3.1.3.6. *Derivation in the time-domain*

A causal function can appear in the following form:

$$g(t)=f(t)u(t)$$

If $F(p)=\mathcal{L}\{g(t)\}$:

– derivation of $f(t)$:

$$\boxed{pF(p)-f(0)=\mathcal{L}\{f'(t)\}} \quad [3.8]$$

By generalizing this we obtain:

$$\boxed{p^n F(p) - p^{n-1} f(0) - p^{n-2} f'(0) - ..... - f^{(n-1)}(0) = \mathcal{L}\left\{f^{(n)}(t)\right\}} \quad [3.9]$$

– derivation of $g(t)$:

$$\boxed{pF(p) = \mathcal{L}\left\{g'(t)\right\}} \quad [3.10]$$

By generalizing this we obtain:

$$\boxed{p^n F(p) = \mathcal{L}\left\{g^{(n)}(t)\right\}} \quad [3.11]$$

3.1.3.7. *Integration in the frequency-domain*

If $F(p) = \mathcal{L}\left\{f(t)\right\}$:

$$\boxed{\int_p^{\infty} F(x)dx = \mathcal{L}\left\{\frac{f(t)}{t}\right\}} \quad [3.12]$$

3.1.3.8. *Integration in the time-domain*

If $F(p) = \mathcal{L}\left\{f(t)\right\}$:

– $f(t)$ has no impulse at its origin $t = 0$:

$$\boxed{\frac{F(p)}{p} = \mathcal{L}\left\{\int_0^t f(x)dx\right\}} \quad [3.13]$$

– $f(t)$ has an impulse with a weight of a at its origin $t = 0$:

$$f(t) = f_1(t) + a\delta(t)$$

$$\boxed{\frac{F_1(p) + a}{p} = \mathcal{L}\left\{\int_0^t f(x)dx\right\}} \quad [3.14]$$

3.1.3.9. *Time-convolution*

If $F_1(p)=\mathcal{L}\{f_1(t)\}$ and $F_2(p)=\mathcal{L}\{f_2(t)\}$:

$$\boxed{F_1(p)F_2(p)=\mathcal{L}\{f_1(t)*f_2(t)\}} \qquad [3.15]$$

with $f_1(t)*f_2(t)=\int_0^t f_1(x)f_2(t-x)dx$

3.1.3.10. *Frequency-convolution*

If $F_1(p)=\mathcal{L}\{f_1(t)\}$ and $F_2(p)=\mathcal{L}\{f_2(t)\}$:

$$\boxed{\frac{1}{j2\pi}F_1(p)*F_2(p)=\mathcal{L}\{f_1(t)f_2(t)\}} \qquad [3.16]$$

with $F_1(p)*F_2(p)=\int_{\sigma-j\infty}^{\sigma+j\infty} F_1(x)F_2(p-x)dx$

3.1.3.11. *Periodic causal signal*

Consider a periodic causal function such that:

$$\begin{cases} f(t+nT)=f(t) & t\geq 0 \\ f(t)=0 & t<0 \end{cases}$$

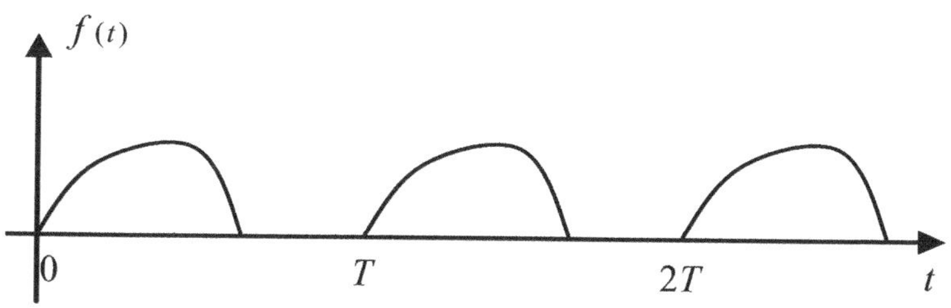

Figure 3.1. *Periodic causal signal*

Let $f_0(t)$ be a function such that:

$$f_0(t)=\begin{cases} f(t) & t\in[0,T] \\ 0 & t<0 \text{ et } t>T \end{cases}$$

We can then write: $f(t)=\sum_{n=0}^{n=\infty} f_0(t-nT)$

if $F_0(p)=\mathcal{L}\{f_0(t)\}$ and $F(p)=\mathcal{L}\{f(t)\}$

Then: $\boxed{F(p)=\dfrac{F_0(p)}{1-e^{-pT}}}$ [3.17]

3.1.4. *Final value and initial value theorems*

3.1.4.1. *Initial value theorem*

Consider $F(p)=\mathcal{L}\{f(t)\}$, then:

$$\boxed{f(0)=\lim_{p\to\infty} pF(p)}$$ [3.18]

3.1.4.2. *Final value theorem*

Consider $F(p)=\mathcal{L}\{f(t)\}$, then:

$$\boxed{f(\infty)=\lim_{p\to 0} pF(p)}$$ [3.19]

3.1.5. *Determining reverse transforms*

3.1.5.1. *Reversal formula*

We use the relation defining the reverse Laplace transform:

$$\boxed{f(t)=\frac{1}{j2\pi}\int_{\sigma-j\infty}^{\sigma+j\infty} F(p)e^{pt}\,dp}$$

Integration is performed in a complex plane most of the time using the residue theorem.

This method is not often used in practice.

3.1.5.2. *Using theorems and tables*

The key is to identify the desired transforms in a given table. There are also times where the properties of the Laplace transform can be used. For example:

Let $F(p)$ be a function such that:

$$F(p)=\frac{4}{p^2+2p+10}=\frac{4}{(p+1)^2+9}$$

If we look in the tables, we find the reverse transform of:

$$F(p)=\frac{\omega_0}{(p+a)^2+\omega_0^2}$$

Which is: $f(t)=e^{-at}\sin\omega_0 t\,u(t)$

We then write: $\boxed{f(t)=\frac{4}{3}e^{-t}\sin 3t\,u(t)}$

3.1.5.3. *Partial fraction expansion*

This method is usable when $F(p)$ is a rational fraction.

$$F(p)=\frac{N(p)}{D(p)}=\frac{a_m p^m+a_{m-1}p^{m-1}\ldots\ldots a_1 p+a_0}{b_n p^n+b_{n-1}p^{n-1}\ldots\ldots b_1 p+b_0}$$

We will suppose that $n>m$, because if $n\leq m$ we return to the previous case by dividing the polynomials.

– Situation with simple poles:

In this case, the rational fraction appears as follows:

$$F(p) = \sum_{i=1}^{i=n} \frac{C_i}{p - p_i} \qquad [3.20]$$

We obtain coefficients C_i by:

$$C_i = (p - p_i) F(p)\big|_{p=p_i} = \frac{N(p_i)}{(p_i - p_1)....(p_i - p_{i-1})(p_i - p_{i+1}).....(p_i - p_n)}$$

Or even: $C_i = \left.\frac{N(p)}{D'(p)}\right|_{p=p_i}$ with $D'(p) = \frac{dD(p)}{dp}$

Looking at the tables we find: $\frac{1}{p+a} = \mathcal{L}\{e^{-at} u(t)\}$

Thus: $$f(t) = \sum_{i=1}^{i=n} C_i e^{p_i t} u(t) \qquad [3.21]$$

– Situation where there are multiple poles:

Take p_0 a multiple pole of order m, the rational fraction can here appear as:

$$F(p) = \frac{N(p)}{(p - p_0)^m D_1(p)}$$

$D_1(p)$ contains only simple poles.

$F(p)$ is written as follows:

$$F(p) = \frac{k_0}{(p - p_0)^m} + \frac{k_1}{(p - p_0)^{m-1}} + ... \frac{k_{m-1}}{(p - p_0)} + \frac{N(p)}{D_1(p)} \qquad [3.22]$$

We obtain coefficients k_i :

$$k_i = \frac{1}{i!} \frac{d^i\left(F(p)(p-p_0)^m\right)}{dp^i}\Bigg|_{p=p_0}$$

The term $\dfrac{N(p)}{D_1(p)}$ contains only simple poles and can be processed the same way as in the previous case.

In the tables we can find: $\dfrac{1}{(p+a)^n} = \mathcal{L}\left\{\dfrac{1}{(n-1)!} t^{n-1} e^{-at} u(t)\right\}$

Thus the result:

$$\boxed{\frac{k_i}{(p\quad p_0)^{m-i}} = \mathcal{L}\left\{\frac{k_i}{(m-i-1)} t^{m-i-1} e^{p_0 t} u(t)\right\}} \qquad [3.23]$$

3.1.6. *Approximation methods*

3.1.6.1. *Step function*

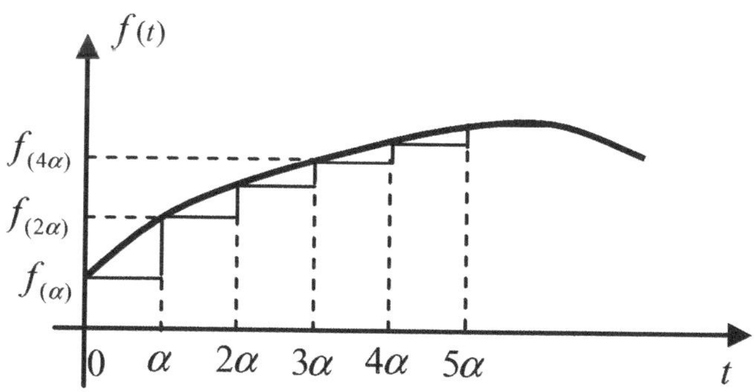

Figure 3.2. *Approximation of a step function*

The approximation of $f(t)$ appears as:

$$f(t) = f(0)u(t) + \big(f(\alpha) - f(0)\big)u(t-\alpha) + ... + \big(f(n\alpha) - f((n-1)\alpha)\big)u(t-n\alpha) + ...$$

$$= \sum_{n=0}^{n=\infty} \big(f(n\alpha) - f((n-1)\alpha)\big)u(t-n\alpha)$$

The Laplace transform is then written as follows:

$$\boxed{F(p) = \frac{1-e^{-\alpha p}}{p} \sum_{n=0}^{n=\infty} f(n\alpha)e^{-n\alpha p}} \qquad [3.24]$$

If p is sufficiently small: $F(p) = \alpha \sum_{n=0}^{n=\infty} f(n\alpha)e^{-n\alpha p}$

3.1.6.2. *Approximation using line segments*

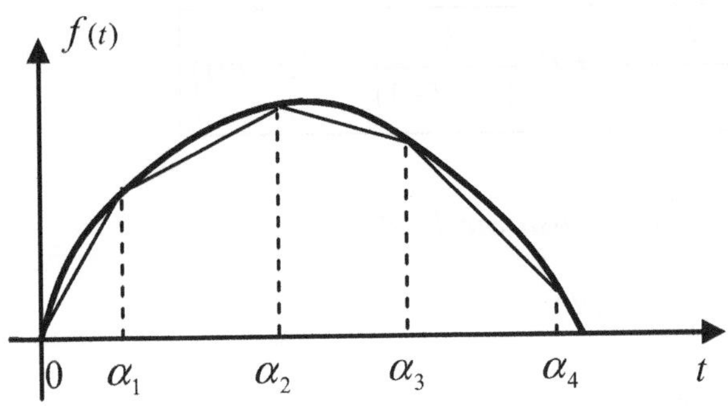

Figure 3.3. *Approximation using line segments*

We derive twice to obtain a Dirac series.

There are two possible approaches:

– we consider that the function will use the following form $f(t)u(t)$ and we apply the time-domain derivation property:

$$f''(t)u(t) = \mathcal{L}^{-1}\left\{p^2F(p) - pf(0) - f'(0)\right\}$$

The second derivative then appears:

$$f''(t)=\sum_{n=1}^{n=\infty} a_n\delta(t-\alpha_n)$$

And we can then find:

$$F(p)=\frac{f'(0)}{p^2}+\frac{f(0)}{p}+\frac{1}{p^2}\sum_{n=1}^{n=\infty} a_n e^{-\alpha_n p} \qquad [3.25]$$

– we can directly consider $f(t)$ and set the second derivative under the following form: $f''(t)=\sum_{n=1}^{n=\infty} a_n\delta(t-\alpha_n)$

In this case $f''(t)=\mathcal{L}^{-1}\left\{p^2F(p)\right\}$ as all the initial conditions are included in the origin of the derivative form of $f(t)$

We obtain: $$F(p)=\frac{1}{p^2}\sum_{n=0}^{n=\infty} a_n e^{-\alpha_n p} \qquad [3.26]$$

3.1.7. *Laplace transform and differential equations*

This method allows us to transform a differential equation of order n by an algebraic equation and to directly introduce the initial conditions.

If we have $f(t)=\mathcal{L}^{-1}\left\{F(p)\right\}$

Then: $f^{(n)}(t)=\mathcal{L}^{-1}\left\{p^nF(p)-p^{n-1}f(0)-p^{n-2}f'(0)-...f^{(n-1)}(0)\right\}$

Consider a linear differential equation with constant coefficients:

$$a_nf^{(n)}(t)+a_{n-1}f^{(n-1)}(t)+.......+a_1f'(t)+a_0f(t)=g(t)$$

By using the Laplace transforms of both:

$$\left(a_n p^n + a_{n-1} p^{n-1} + \ldots\ldots + a_1 p + a_0\right) F(p) = G(p) + I(p)$$

with $G(p) = \mathcal{L}\{g(t)\}$ and $I(p)$ depending on the initial conditions.

We then obtain: $$F(p) = \frac{G(p) + I(p)}{a_n p^n + a_{n-1} p^{n-1} + \ldots\ldots + a_1 p + a_0}$$

We obtain $f(t)$ using the reverse transform of $F(p)$.

3.1.8. *Table of common Laplace transforms*

The following table presents the most common Laplace transforms as well as the ROC.

It can be added to by applying the properties specific to Laplace transforms.

$f(t) = \mathcal{L}^{-1}\{F(p)\}$	$F(p) = \mathcal{L}\{f(t)\}$	ROC
$\delta(t)$	1	$\forall p$
$\delta(t-\tau)$	$e^{-\tau p}$	$\forall p$
$u(t)$	$\frac{1}{p}$	$\Re e(p) > 0$
$u(t-\tau)$	$\frac{1}{p} e^{-\tau p}$	$\Re e(p) > 0$
$u(t) - u(t-\tau)$	$\frac{1}{p}\left(1 - e^{-\tau p}\right)$	$\Re e(p) > 0$
$tu(t)$	$\frac{1}{p^2}$	$\Re e(p) > 0$
$\frac{1}{n!} t^n u(t)$	$\frac{1}{p^{n+1}}$ n entier > 0	$\Re e(p) > 0$
$\frac{1}{\Gamma_{(q+1)}} t^q u(t)$	$\frac{1}{p^{q+1}}$	$\Re e(p) > 0$

$e^{-at}u(t)$	$\frac{1}{p+a}$	$\Re e(p) > -a$
$\frac{1}{n!}t^n e^{-at}u(t)$	$\frac{1}{(p+a)^{n+1}}$	$\Re e(p) > -a$
$(1-e^{-at})u(t)$	$\frac{a}{p(p+a)}$	$\Re e(p) > 0$
$\sin \omega t\, u(t)$	$\frac{\omega}{p^2+\omega^2}$	$\Re e(p) > 0$
$\cos \omega t\, u(t)$	$\frac{p}{p^2+\omega^2}$	$\Re e(p) > 0$
$\sinh at\, u(t)$	$\frac{a}{p^2-a^2}$	$\Re e(p) > \|a\|$
$\cosh at\, u(t)$	$\frac{p}{p^2-a^2}$	$\Re e(p) > \|a\|$
$e^{-at}\sin \omega t\, u(t)$	$\frac{\omega}{(p+a)^2+\omega^2}$	$\Re e(p) > -a$
$e^{-at}\cos \omega t\, u(t)$	$\frac{p+a}{(p+a)^2+\omega^2}$	$\Re e(p) > -a$
$\sqrt[n]{t}u(t)$	$\Gamma(1+\frac{1}{n})p^{-\left(1+\frac{1}{n}\right)}$	$\Re e(p) > 0$
$\ln\left(\frac{t}{\tau}\right)u(t)$	$-\frac{\tau}{p}\left(\ln(\tau p)+\gamma\right)$	$\Re e(p) > 0$
$J_n(\omega t)u(t)$	$\frac{\omega^n\left(p+\sqrt{p^2+\omega^2}\right)^{-n}}{\sqrt{p^2+\omega^2}}$	$\Re e(p) > 0$ $n > -1$
$I_n(\omega t)u(t)$	$\frac{\omega^n\left(p+\sqrt{p^2-\omega^2}\right)^{-n}}{\sqrt{p^2-\omega^2}}$	$\Re e(p) > \|\omega\|$
$erf(t)u(t)$	$\frac{e^{\frac{p^2}{4}} erfc(\frac{p}{2})}{p}$	$\Re e(p) > 0$

Table 3.1. *Laplace transform of common function*

3.1.9. *Transient state and steady state*

In the case of a linear system designed using lumped elements, the transfer function appears as a rational fraction.

$$H(p) = \frac{N(p)}{D(p)} = \frac{N(p)}{(p-p_1)(p-p_2)\ldots\ldots.(p-p_n)}$$

where $p_1, p_2, \ldots\ldots, p_n$ are the poles of $H(p)$

The Laplace transform of the response to system $s(t)$ appears as follows:

$$S(p) = H(p)E(p)$$

$E(p)$ being the Laplace transform of input signal $e(t)$:

$$S(p) = \frac{N(p)E(p)}{(p-p_1)(p-p_2)\ldots\ldots.(p-p_n)}$$

This can be expanded into partial fractions:

$$S(p) = \frac{C_1}{p-p_1} + \frac{C_2}{p-p_2} + \ldots. + \frac{C_n}{p-p_n} + \Phi(p)$$

$\Phi(p)$ is dependant on the input signal.

We write $s_1(t)$ the reverse transform of $\Phi(p)$:

$$s(t) = \sum_{i=1}^{i=n} C_i e^{p_i t} + s_1(t)$$

The response contains two components:

– the transient state which is characteristic of the linear system:

$$\sum_{i=1}^{i=n} C_i e^{p_i t}$$

– the steady state which depends on the input signal:

$s_1(t)$

The transient state is also called the natural response.

The steady state is also called the forced response.

In the case of stable linear systems, we demonstrate that all the poles of $H(p)$ have negative real parts. From this, we establish that the transient state approaches zero when time increases to infinity.

3.2. Exercise instruction

The exercises presented in this chapter come from B.P. Lathi [LAT 66].

3.2.1. *Exercise 3.1*

Determine the region of convergence (ROC) of the following Laplace transforms:

$f_1(t) = e^t u(t)$ $\qquad f_2(t) = e^{-t} u(t)$ $\qquad f_3(t) = \sin\omega_0 t\, u(t)$

$f_4(t) = te^{-at} u(t)$ $\qquad f_5(t) = \dfrac{1}{1+t^2} u(t)$ $\qquad f_6(t) = \dfrac{1}{1+e^t} u(t)$

3.2.2. *Exercise 3.2*

Determine the Laplace transforms of the following functions:

$f_1(t) = u(t) - u(t-1)$ $\qquad f_2(t) = t\sin\omega_0 t\, u(t)$

$f_3(t) = (t-\tau)\sin\omega_0(t-\tau)u(t-\tau)$ $\qquad f_4(t) = \dfrac{1-\cos\omega_0 t}{t} u(t)$

$f_5(t) = t^2 e^{-t} u(t)$ $\qquad f_6(t) = t^2 e^{-t} u(t-\tau)$

$f_7(t) = \sin\omega_1 t \cos\omega_2 t u(t)$ $\qquad f_8(t) = e^{-t}\displaystyle\int_0^t \tau\sin\omega_0\tau\, d\tau$

$f_9(t) = \dfrac{e^{-at}\sin\omega_0 t}{t} u(t)$ $\qquad f_{10}(t) = te^{-at}\cos\omega_0 t u(t)$

$f_{11}(t) = \dfrac{e^{-t}-1}{t} u(t)$

3.2.3. *Exercise 3.3*

Compare the Laplace transforms of the following functions:

$$f_1(t) = \sin \omega_0 t u(t) \qquad f_2(t) = \sin \omega_0 (t-\tau) u(t-\tau)$$
$$f_3(t) = \sin \omega_0 (t-\tau) u(t) \qquad f_4(t) = \sin \omega_0 (t) u(t-\tau)$$

3.2.4. *Exercise 3.4*

Determine the reverse transforms of the following functions:

$$F_1(p) = \frac{10p}{p^2+25} \qquad F_2(p) = \frac{2p-10}{p^2+25} \qquad F_3(p) = \frac{p^2+2p+2+\frac{6}{p}}{p^2}$$

$$F_4(p) = \frac{p^2+4p+4}{p^2+2p+2} \qquad F_5(p) = \frac{p}{(p+a)^2} \qquad F_6(p) = \frac{25p^2+10p+2}{p^3}e^{-5p}$$

$$F_7(p) = 20(p+\alpha)^{-3} \qquad F_8(p) = \frac{e^{-p}+e^{-2p}}{(p+1)(p+2)} \qquad F_9(p) = \frac{e^{-(p-1)}}{(p-1)^2+4}$$

$$F_{10}(p) = \frac{pe^{-p}}{p+a} \qquad F_{11}(p) = \frac{p^2}{(p+a)^2+b^2} \qquad F_{12}(p) = \ln\left(\frac{p+1}{p+2}\right)$$

$$F_{13}(p) = p\ln\left(\frac{p+1}{p+2}\right) \qquad F_{14}(p) = \frac{1}{p}\ln\left(\frac{p+1}{p+2}\right) \qquad F_{15}(p) = \frac{(T(p+1)+1)e^{-T(p+\alpha)}}{p^2+2\alpha p+\alpha^2}$$

$$F_{16}(p) = \frac{2p^2+5p+9}{(p+1)^2(p+7)} \qquad F_{17}(p) = \frac{p-1+e^{-p}}{p^2(1-e^{-p})}$$

3.2.5. *Exercise 3.5*

Let $f(t)$ be a function with the following Laplace transform:

$$F(p) = \frac{6p^2+8p-5}{p(2p^2+6p+5)}$$

Determine $f(0)$ and $f(\infty)$, the initial and final values of $f(t)$.

3.2.6. *Exercise 3.6*

Let us consider the following functions:

$$f_1(t) = e^{-\alpha t} u(t) \quad f_2(t) = \left(1 - e^{-\alpha t}\right) u(t-\tau) \quad f_3(t) = t e^{-\alpha t} u(t)$$

– Determine the Laplace transforms $F_1(p)$, $F_2(p)$ and $F_3(p)$.

– From these results, find initial and final values $f_1(t)$, $f_2(t)$ as well as $f_3(t)$.

– We note $A(t) = \int_0^t f(\tau) d\tau$. Demonstrate that the area contained under graph $f(t)$ can be written $A = A(\infty) = \lim_{p \to 0} F(p)$. Apply this result to the previous functions.

3.2.7. *Exercise 3.7*

Determine the Laplace transforms of the following functions:

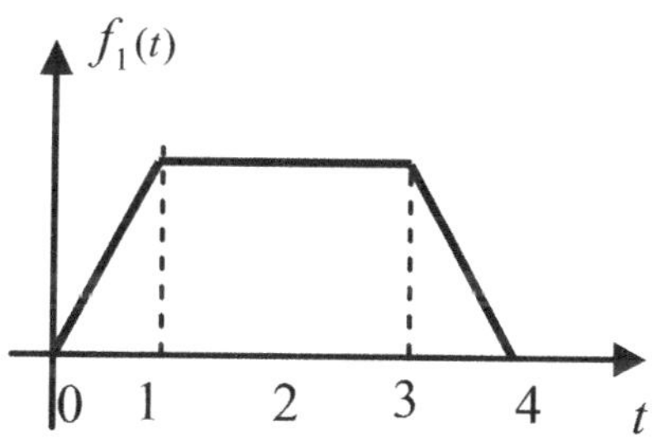

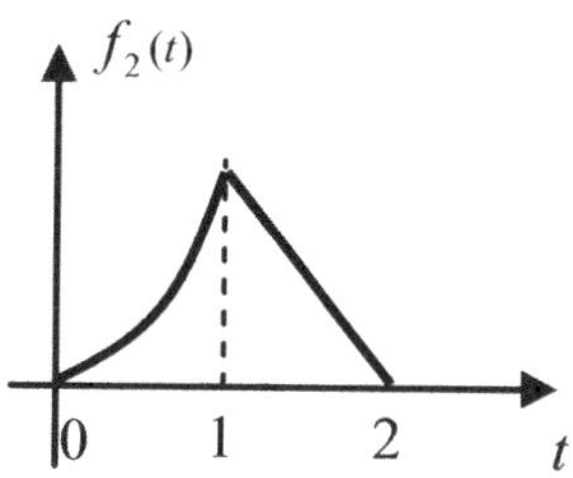

$$\begin{cases} f_2(t) = 0 & t < 0 \\ f_2(t) = t^2 & t \in [0,1] \\ f_2(t) = 2 - t & t \in [1,2] \\ f_2(t) = 0 & t > 2 \end{cases}$$

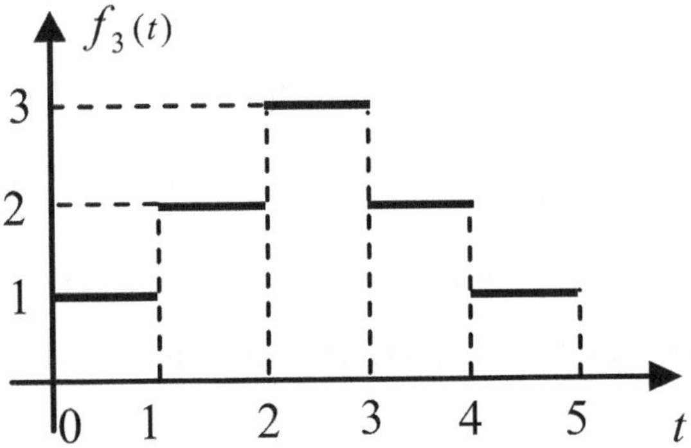

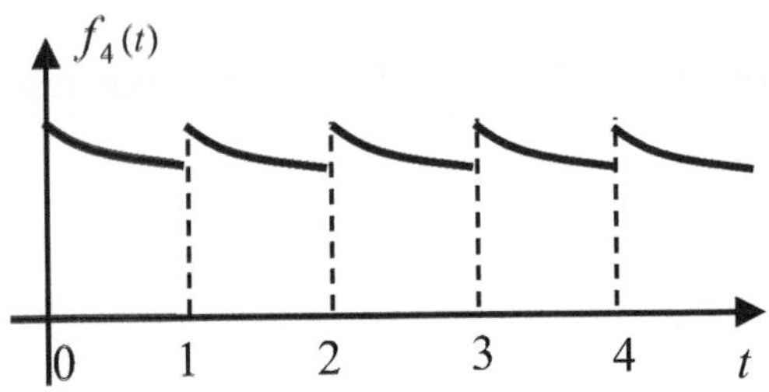

$$f_4(t) = e^{-2t} \qquad t \in [0,1]$$

period $T = 1$

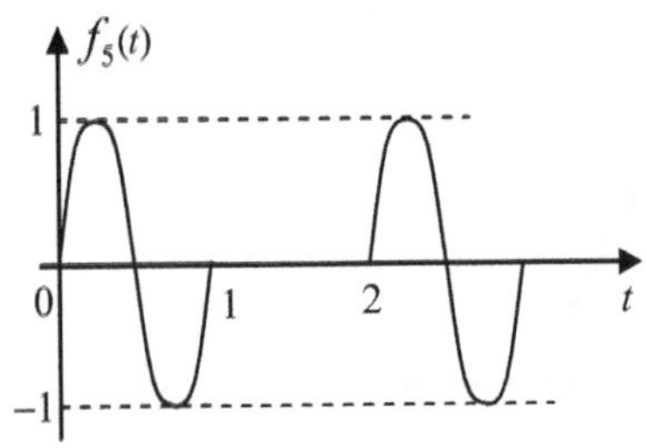

$$f_5(t) = \begin{cases} \sin 2\pi t & \text{where } t \in [0,1] \\ 0 & \text{where } t \in [1,2] \end{cases}$$

period $T = 2$

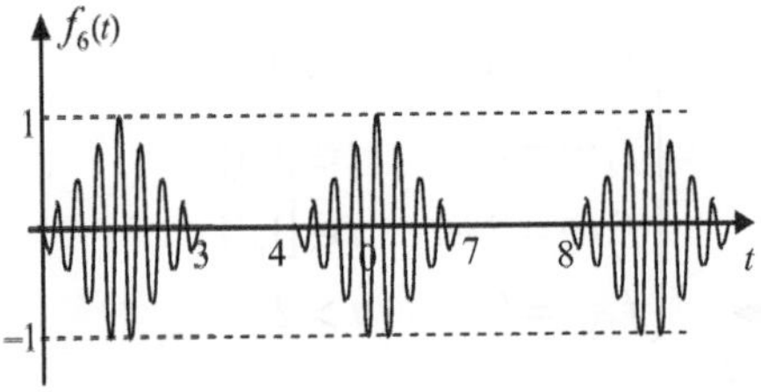

$$f_6(t) = \begin{cases} \frac{2}{3} t \sin 200\pi t & t \in \left[0, \frac{3}{2}\right] \\ \left(-\frac{2}{3} t + 2\right) \sin 200\pi t & t \in \left[\frac{3}{2}, 3\right] \\ 0 & t \in [3,4] \end{cases}$$

3.2.8. *Exercise 3.8*

Solve the following differential equations applying the Laplace transform:

$y'' + 4y' + 4y = 2e^{-x}$ initial conditions: $y(0) = 2 \;\; y'(0) = 5$

$y^{(4)} + 4y^{(3)} + 8y''$

$+ 8y' + 3y + e^{-x} = 0$ null initial conditions

$y^{(3)} + 5y'' + 8y' + 4y = e^{-x}$ initial conditions: $y(0) = y'(0) = 2 \;\; y''(0) = 1$

3.2.9. *Exercise 3.9*

Solve the following differential equations systems:

$$\begin{cases} y_1'' + 2y_2' + 2y_1 - 2y_2 = 10u(x) \\ y_2'' + 2y_1' + 2y_2 - 2y_1 = 0 \end{cases}$$ null initial conditions

$$\begin{cases} y_1'' + y_1' + y_1 - y_2 = u(x) - u(x-2) \\ 2y_2' + y_2 - y_1 = 0 \end{cases}$$ null initial conditions

3.2.10. *Exercise 3.10*

Let $e(t)$ be the following periodic voltage:

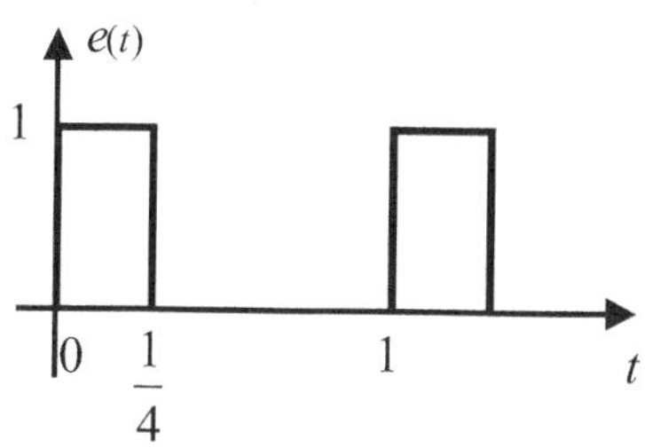

$$e(t) = \begin{cases} 1 & t \in \left[0, \frac{1}{4}\right] \\ 0 & t \in \left[\frac{1}{4}, 1\right] \end{cases}$$

period $T = 1$

We apply this voltage to the terminals of an RLC series circuit

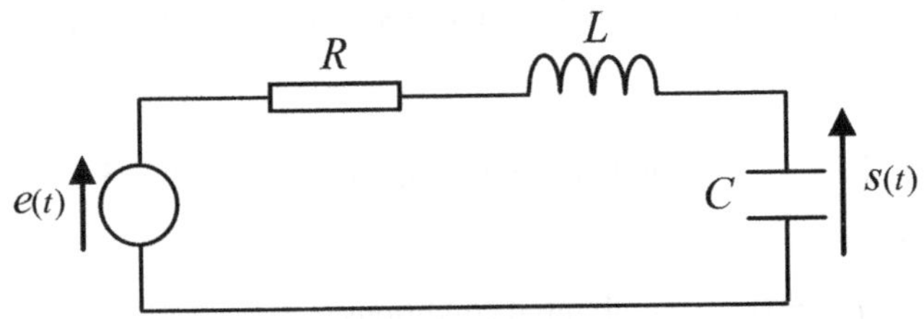

$R = 3\Omega$

$L = 1H$

$C = \frac{1}{2}F$

Determine $s(t)$ and draw its graph.

3.3. Solutions to the exercises

3.3.1. *Exercise 3.1*

In order to determine the ROC, we need to establish α such that:

$$\lim_{t\to\infty} e^{-\alpha t} f(t) = 0$$

$$\lim_{t\to\infty} e^{-\alpha t} f_1(t) = \lim_{t\to\infty} e^{-\alpha t} e^{t} u(t) = \lim_{t\to\infty} e^{(1-\alpha)t}$$

This limit is null if $(1-\alpha) < 0$ therefore if $\boxed{\alpha > 1}$

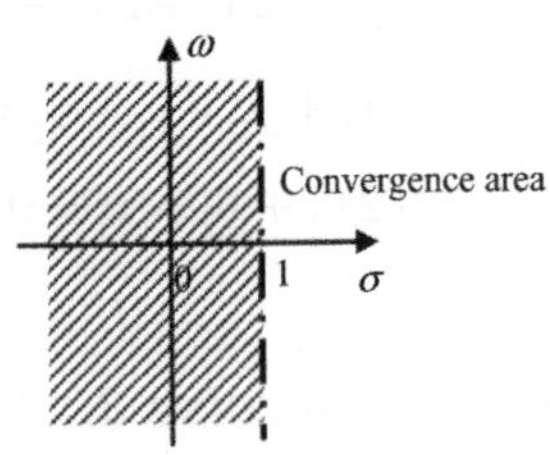

– $f_2(t) = e^{-t} u(t) \quad \Rightarrow \alpha > -1$

– $f_3(t) = \sin \omega_0 t u(t) \quad \Rightarrow \alpha > 0$

– $f_4(t) = t e^{-at} u(t) \quad \Rightarrow \alpha > -a$

– $f_5(t) = \dfrac{1}{1+t^2} u(t) \quad \Rightarrow \alpha > 0$

– $f_6(t) = \dfrac{1}{1+e^{t}} u(t) \quad \Rightarrow \alpha > -1$

3.3.2. *Exercise 3.2*

– $f_1(t) = u(t) - u(t-1)$:

$\mathcal{L}\{u(t)\} = \dfrac{1}{p}$, we apply the property of time-domain translation:

$$\mathcal{L}\{u(t-1)\} = \frac{e^{-p}}{p}$$

$$\boxed{F_1(p) = \frac{1}{p}\left(1 - e^{-p}\right)}$$ exists no matter the value of p

– $f_2(t) = t \sin \omega_0 t \ u(t)$:

$\mathcal{L}\{\sin \omega_0 t \ u(t)\} = \dfrac{\omega_0}{p^2 + \omega_0^2}$, we apply the property of frequency-domain derivation:

$$\mathcal{L}\{-tf(t)\} = \frac{dF(p)}{dp}$$

Thus: $\boxed{F_2(p) = \dfrac{2\omega_0 p}{\left(p^2 + \omega_0^2\right)^2}}$ ROC $\Re e(p) > 0$

– $f_3(t) = (t - \tau)\sin \omega_0 (t - \tau) \ u(t - \tau)$:

$f_3(t) = f_2(t - \tau)$, we apply the property of time-domain translation:

$$\mathcal{L}\{f(t-\tau)\} = F(p)e^{-p\tau}$$

Thus: $\boxed{F_3(p) = \dfrac{2\omega_0 p e^{-p\tau}}{\left(p^2 + \omega_0^2\right)^2}}$ ROC $\Re e(p) > 0$

– $f_4(t) = \dfrac{1-\cos\omega_0 t}{t}\ u(t)$:

$\mathcal{L}\{(1-\cos\omega_0 t)\ u(t)\} = \dfrac{1}{p} - \dfrac{p}{p^2+\omega_0^2} = \dfrac{\omega_0^2}{p(p^2+\omega_0^2)}$, we apply the property of frequency-domain integration:

$$\mathcal{L}\left\{\frac{f(t)}{t}\right\} = \int_p^{\infty} F(x)dx$$

Thus: $\boxed{F_4(p) = \ln\left(\sqrt{1+\dfrac{\omega_0^2}{p^2}}\right)}$ ROC $\Re e(p) > 0$

– $f_5(t) = t^2 e^{-t}\ u(t)$:

$\mathcal{L}\{e^{-t}\ u(t)\} = \dfrac{1}{p+1}$, we apply the property of frequency-domain derivation:

$$\mathcal{L}\{t^2 f(t)\} = \frac{d^2F(p)}{dp^2}$$

Thus: $\boxed{F_5(p) = \dfrac{2}{(p+1)^3}}$ ROC $\Re e(p) > -1$

– $f_6(t) = t^2 e^{-t}\ u(t-\tau)$:

$$F_6(p) = \int_0^{\infty} t^2 e^{-t}\ u(t-\tau) e^{-pt} dt = \int_\tau^{\infty} t^2 e^{-(p+1)t} dt,$$

When we integrate in parts we obtain:

$$\boxed{F_6(p) = \left(\frac{\tau^2}{p+1} + \frac{2\tau}{(p+1)^2} + \frac{2}{(p+1)^3}\right) e^{-\tau(p+1)}}$$

ROC $\Re e(p) > -1$

– $f_7(t) = \sin\omega_1 t \cos\omega_2 t\ u(t)$:

$$\mathcal{L}\{\sin\omega_1 t\ u(t)\} = \frac{\omega_1}{p^2+\omega_1^2}$$

We use the fact that $\cos\omega_2 t = \dfrac{e^{j\omega_2 t}+e^{-j\omega_2 t}}{2}$

And the property of frequency-domain translation $\mathcal{L}\{e^{p_0 t} f(t)\} = F(p-p_0)$

We obtain: $F_7(p) = \dfrac{1}{2}\left(F(p+j\omega_2)+F(p-j\omega_2)\right)$

We replace:

$$\boxed{F_7(p) = \frac{\omega_1}{2}\left(\frac{1}{p^2-2j\omega_2 p+\omega_1^2-\omega_2^2}+\frac{1}{p^2+2j\omega_2 p+\omega_1^2-\omega_2^2}\right)}$$

ROC $\Re e(p) > 0$

– $f_8(t) = e^{-t}\int_0^t \tau\sin\omega_0\tau\ d\tau$:

$$\mathcal{L}\{t\sin\omega_0 t\ u(t)\} = \frac{2\omega_0 p}{\left(p^2+\omega_0^2\right)^2}$$

We apply the properties of time-domain integration and frequency-domain translation:

$$\mathcal{L}\left\{\int_0^t f(u)du\right\} = \frac{F(p)}{p} \text{ and } \mathcal{L}\{e^{-at}f(t)\} = F(p+a)$$

We obtain: $\boxed{F_8(p) = \dfrac{2\omega_0}{\left((p+1)^2+\omega_0\right)^2}}$ ROC $\Re e(p) > -1$

– $f_9(t)=\dfrac{e^{-at}\sin\omega_0 t}{t}u(t)$:

$$\mathcal{L}\{\sin\omega_0 t\ u(t)\}=\frac{\omega_0}{p^2+\omega_0^2}$$

We apply the properties of integration and translation in the frequency domain:

$$\mathcal{L}\{e^{-at}f(t)\}=F(p+a)\text{ and }\mathcal{L}\left\{\frac{f(t)}{t}\right\}=\int_p^\infty F(x)dx$$

We obtain: $\boxed{F_9(p)=\dfrac{\pi}{2}-Arctg\dfrac{p+a}{\omega_0}}$ ROC $\Re e(p)>-a$

– $f_{10(t)}=t\,e^{-at}\cos\omega_0 t\ u(t)$:

$$\mathcal{L}\{\cos\omega_0 t\ u(t)\}=\frac{p}{p^2+\omega_0^2}$$

We apply the properties of translation and derivation in the frequency-domain:

$$\mathcal{L}\{e^{-at}f(t)\}=F(p+a)\text{ and }\mathcal{L}\{-tf(t)\}=\frac{F(p)}{dp}$$

We obtain: $\boxed{F_{10}(p)=\dfrac{(p+a)^2-\omega_0^2}{\left((p+a)^2+\omega_0^2\right)^2}}$ ROC $\Re e(p)>-a$

– $f_{11}(t)=\dfrac{e^{-t}-1}{t}u(t)$:

$$\mathcal{L}\{(e^{-t}-1)\,u(t)\}=\frac{1}{p+1}-\frac{1}{p}$$

We apply the property of frequency-domain integration:

$$\mathcal{L}\left\{\frac{f(t)}{t}\right\} = \int_{p}^{\infty} F(x)dx$$

We obtain: $\boxed{F_{11}(p) = \ln\frac{p}{p+1}}$ ROC $\Re e(p) > -1$

3.3.3. *Exercise 3.3*

– $f_1(t) = \sin\omega_0 t\ u(t)$:

In the tables we obtain:

$$\boxed{F_1(p) = \frac{\omega_0}{p^2+\omega_0^2}}$$

– $f_2(t) = \sin\omega_0(t-\tau)\ u(t-\tau)$:

We write: $f_2(t) = f_2(t-\tau)$. We apply the property of time-domain translation:

$$\boxed{F_2(p) = \frac{\omega_0 e^{-p\tau}}{p^2+\omega_0^2}}$$

– $f_3(t) = \sin\omega_0(t-\tau)\ u(t)$:

We write: $f_3(t) = \left(\cos\omega_0\tau\sin\omega_0 t - \sin\omega_0\tau\cos\omega_0 t\right)u(t)$

Hence: $\boxed{F_3(p) = \frac{\omega_0\cos\omega_0\tau - p\sin\omega_0\tau}{p^2+\omega_0^2}}$

– $f_4(t) = \sin\omega_0 t \; u(t-\tau)$:

We write: $f_4(t) = -f_3(t-\tau)$.We apply the property of time-domain translation

$$\boxed{F_4(p) = \frac{p\sin\omega_0\tau - \omega_0\cos\omega_0\tau}{p^2+\omega_0^2} e^{-p\tau}}$$

$f_1(t)$ and $f_2(t)$ are identical functions translated from τ. So are $f_3(t)$ and $f_4(t)$ only with a different sign.

3.3.4. *Exercise 3.4*

Calculations for the reverse transforms.

– $F_1(p) = \dfrac{10p}{p^2+25}$:

By referring to the tables, we find: $\mathcal{L}\{\cos\omega_0 t \; u(t)\} = \dfrac{p}{p^2+\omega_0^2}$

Hence: $\boxed{f_1(t) = 10\cos 5t \; u(t)}$

– $F_2(p) = \dfrac{2p-10}{p^2+25}$:

By referring to the tables, we find:

$\mathcal{L}\{\sin\omega_0 t \; u(t)\} = \dfrac{\omega_0}{p^2+\omega_0^2}$ and $\mathcal{L}\{\cos\omega_0 t \; u(t)\} = \dfrac{p}{p^2+\omega_0^2}$

Hence: $\boxed{f_2(t) = 2(\cos 5t - \sin 5t)\; u(t) = 2\sqrt{2}\cos\left(5t+\frac{\pi}{4}\right) u(t)}$

– $F_3(p) = \dfrac{p^2 + 2p + 2 + \dfrac{6}{p}}{p^2}$:

We decompose: $F_3(p) = 1 + \dfrac{2}{p} + \dfrac{2}{p^2} + \dfrac{6}{p^3}$

By referring to the tables, we find:

$$\boxed{f_3(t) = \delta(t) + \left(2 + 2t + 3t^2\right)u(t)}$$

– $F_4(p) = \dfrac{p^2 + 4p + 4}{p^2 + 2p + 2}$:

We write: $F_4(p) = 1 + \dfrac{2(p+1)}{p^2 + 2p + 2} = 1 + \dfrac{2(p+1)}{(p+1)^2 + 1}$

By referring to the tables, we obtain:

$$\mathcal{L}\{\cos\omega_0 t\ u(t)\} = \frac{p}{p^2 + \omega_0^2} \text{ and } \mathcal{L}\{e^{-at}\ f(t)\} = F(p+a)$$

Hence: $\mathcal{L}^{-1}\left\{\dfrac{p+1}{(p+1)^2 + 1}\right\} = e^{-t}\cos t\ u(t)$

Thus: $\boxed{f_4(t) = \delta(t) + 2e^{-t}\cos t\ u(t)}$

– $F_5(p) = \dfrac{p}{(p+a)^2}$:

By referring to the tables, we obtain:

$$\mathcal{L}\{te^{-at}\ u(t)\} = \frac{1}{(p+a)^2} \text{ and } \mathcal{L}\{f'(t)\ u(t)\} = pF_{(p)} - f_{(0)}$$

Thus: $\mathcal{L}\left\{\left(te^{-at}\right)' u(t)\right\} = \dfrac{p}{(p+a)^2}$

Hence: $\boxed{f_5(t) = (1-at)e^{-at}\ u(t)}$

– $F_6(p) = \dfrac{25p^2 + 10p + 2}{p^3} e^{-5p}$:

We write: $F_6(p) = \left(\dfrac{25}{p} + \dfrac{10}{p^2} + \dfrac{2}{p^3}\right) e^{-5p}$

We then apply the results:

$$\mathcal{L}\{f(t-\tau)\} = F(p)e^{-p\tau} \text{ and } \mathcal{L}\left\{\left(25 + 10t + t^2\right) u(t)\right\} = \frac{25}{p} + \frac{10}{p^2} + \frac{2}{p^3}$$

Hence: $\boxed{\begin{array}{l} f_6(t) = \left(25 + 10(t-5) + (t-5)^2\right) u(t-5) \\ f_{6(t)} = t^2\ u(t-5) \end{array}}$

– $F_7(p) = 20(p+a)^{-3} = \dfrac{20}{(p+a)^3}$:

We apply the results:

$$\mathcal{L}\left\{e^{-at} f(t)\right\} = F(p+a) \text{ and } \mathcal{L}\left\{\frac{t^2}{2} u(t)\right\} = \frac{1}{p^3}$$

Hence: $\boxed{f_7(t) = 10t^2 e^{-at}\ u(t)}$

– $F_8(p) = \dfrac{e^{-p} + e^{-2p}}{(p+1)(p+2)}$:

We write: $F_8(p) = \dfrac{e^{-p}}{(p+1)(p+2)} + \dfrac{e^{-2p}}{(p+1)(p+2)}$

We use the property of time-domain translation $\mathcal{L}\{f(t-\tau)\} = F(p)e^{-p\tau}$ a partial-fraction expansion:

$$\frac{1}{(p+1)(p+2)} = \frac{1}{p+1} - \frac{1}{p+2}$$

Hence the inverse transform:

$$\mathcal{L}^{1}\left\{\frac{1}{p+1} - \frac{1}{p+2}\right\} = \left(e^{-t} - e^{-2t}\right) u(t)$$

Hence: $\boxed{f_8(t) = \left(e^{-(t-1)} - e^{-2(t-1)}\right) u(t-1) + \left(e^{-(t-2)} - e^{-2(t-2)}\right) u(t-2)}$

In this example, we can show the direct calculation by applying the residue theorem.

$$f(t) = \frac{1}{j2\pi} \int_{\sigma-j\infty}^{\sigma+j\infty} F(p)e^{pt} dp = \sum Residue\left(F(p)e^{pt}\right)$$

Thus

$$f(t) = \sum Residue\left(\frac{e^{-p}}{p+1} - \frac{e^{-p}}{p+2} + \frac{e^{-2p}}{p+1} - \frac{e^{-2p}}{p+2}\right)e^{pt}$$

$$= \sum Residue\left(\frac{e^{-p(1-t)}}{p+1} - \frac{e^{-p(1-t)}}{p+2} + \frac{e^{\ p(2-t)}}{p+1} - \frac{e^{-p(2-t)}}{p+1}\right)$$

The poles are $p_1 = -1$ and $p_2 = -2$, the integration outline must therefore circle these two points with $\Re e(p) < 0$.

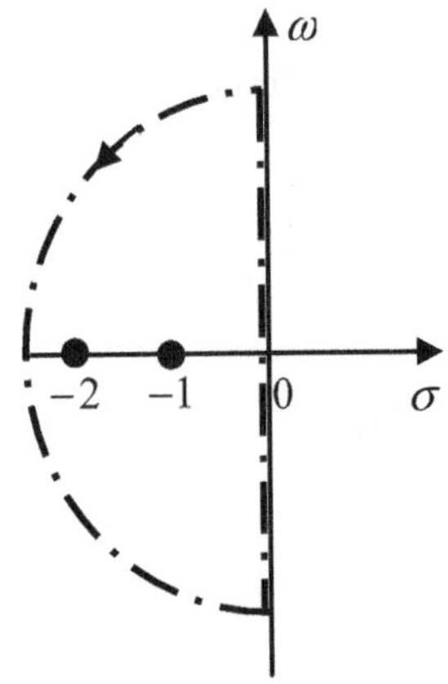

For the first two residue, convergence is assured if $\Re e(-p(1-t))<0$:

$$-\,Residue\left(\frac{e^{-p(1-t)}}{p+1}\right)=\left.\frac{e^{-p(1-t)}}{p+1}\right|_{p=-1}=e^{1-t}$$ for $t>1$, that we can also write as $e^{-(t-1)}u(t-1)$.

$$-\,Residue\left(\frac{e^{-p(1-t)}}{p+2}\right)=\left.\frac{e^{-p(1-t)}}{p+2}\right|_{p=-2}=e^{2(1-t)}$$ for $t>1$, that we can also write as $e^{-2(t-1)}u(t-1)$.

For the two remaining residue, convergence is assured if $\Re e(-p(2-t))<0\mathrm{M}$:

$$-\,Residue\left(\frac{e^{-p(2-t)}}{p+1}\right)=\left.\frac{e^{-p(2-t)}}{p+1}\right|_{p=-1}=e^{2-t}$$ for $t>2$, that we can also write $e^{-(t-2)}u(t-2)$.

$$-\,Residue\left(\frac{e^{-p(2-t)}}{p+2}\right)=\left.\frac{e^{-p(2-t)}}{p+2}\right|_{p=-2}=e^{2(2-t)}$$ for $t>2$, that we can also write $e^{-2(t-2)}u(t-2)$.

We therefore find the previous result.

– $F_9(p) = \dfrac{e^{-(p-1)}}{(p-1)^2+4}$:

We apply the results:

$$\mathcal{L}\{f(t-\tau)\} = F(p)e^{-p\tau}, \ \mathcal{L}\{e^{-at} f(t)\} = F(p+a)$$

$$\text{and } \mathcal{L}\{\sin 2t \ u(t)\} = \frac{2}{p^2+4}$$

We obtain: $\boxed{f_9(t) = \frac{1}{2} e^t \sin 2(t-1) \ u(t-1)}$

– $F_{10}(p) = \dfrac{pe^{-p}}{p+a}$:

We apply the results:

$$\mathcal{L}\{f(t-\tau)\} = F(p)e^{-p\tau}, \mathcal{L}\{e^{-at} u(t)\} = \frac{1}{p+a} \text{ et}$$

$$\mathcal{L}\{f'(t) \ u(t)\} = pF(p) - f(0)$$

Hence: $\mathcal{L}\left\{\left(e^{-at}\right)' + \delta(t)\right\} = \dfrac{p}{p+a}$

We obtain: $\boxed{f_{10}(t) = -ae^{-a(t-1)} \ u(t-1) + \delta(t-1)}$

– $F_{11}(p) = \dfrac{p^2}{(p+a)^2+b^2}$:

We apply the results:

$$\mathcal{L}\{e^{-at} \sin bt \ u(t)\} = \frac{b}{(p+a)^2+b^2}$$

$$\text{and } \mathcal{L}\{f'(t) \ u(t)\} = pF(p) - f(0)$$

Hence:
$$\frac{p}{(p+a)^2+b^2}=\mathcal{L}^{1}\left\{\frac{1}{b}\left(e^{-at}\sin bt\right)' u(t)\right\}$$
$$=\mathcal{L}^{1}\left\{\frac{e^{-at}}{b}(b\cos bt-a\sin bt)\,u(t)\right\}$$

By applying the derivation property a second time, we obtain:

$$\boxed{f_{11}(t)=\delta(t)+\frac{e^{-at}}{b}\left(\left(a^2-b^2\right)\sin bt-2ab\cos bt\right)u(t)}$$

– $F_{12}(p)=\ln\dfrac{p+1}{p+2}$:

We write: $F_{12(p)}=\left[\ln\dfrac{x+1}{x+2}\right]_p^{\infty}=\int_p^{\infty}\left(\dfrac{1}{x+2}-\dfrac{1}{x+1}\right)dx$

We apply the results:

$$\mathcal{L}\left\{\frac{f(t)}{t}\right\}=\int_p^{\infty}F(p)dp$$

Hence: $\boxed{f_{12}(t)=\dfrac{e^{-2t}-e^{-t}}{t}u(t)}$

– $F_{13}(p)=p\ln\dfrac{p+1}{p+2}$:

We apply the result:

$$\mathcal{L}\left\{f'(t)+f(0)\delta(t)\right\}=pF(p)\text{ with }f(0)=-1$$

Hence: $\boxed{f_{13}(t)=\dfrac{(t+1)e^{-t}-(2t+1)e^{-2t}}{t^2}u(t)-\delta(t)}$

$$F_{14}(p)=\frac{1}{p}\ln\frac{p+1}{p+2}$$

We apply the result:

$$\mathcal{L}\left\{\int_0^t f(x)dx\right\}=\frac{F(p)}{p}$$

Hence: $\boxed{f_{14}(t)=\int_0^t \frac{e^{-2x}-e^{-x}}{x}dx}$

– $F_{15}(p)=\dfrac{\left(T(p+\alpha)+1\right)e^{-T(p+\alpha)}}{p^2+2\alpha p+\alpha^2}$:

We write: $F_{15}(p)=\dfrac{T(p+\alpha)+1}{(p+\alpha)^2}e^{-T(p+\alpha)}$

We apply the result:

$$\mathcal{L}\{f(t-T)\}=F(p)e^{-pT},\ \mathcal{L}\left\{f(t)e^{-\alpha t}\right\}=F(p+\alpha)$$

and $\mathcal{L}\{t\,u(t)\}=\dfrac{1}{p^2}$

Hence: $\boxed{\begin{aligned}f_{15}(t)&=\left(t-T\right)e^{-\alpha(t-T)}u(t-T)+Te^{-\alpha(t-T)}u(t-T)\\&=te^{-\alpha(t-T)}u(t-T)\end{aligned}}$

– $F_{16}(p)=\dfrac{2p^2+5p+9}{(p+1)^2(p+7)}$:

We expand into partial-fractions:

$$F_{16}(p)=\frac{A}{p+7}+\frac{K_0}{(p+1)^2}+\frac{K_1}{p+1}$$

with $A=1 \quad K_0=1 \quad K_1=0$

Hence: $\boxed{f_{16}(t) = \left(2e^{-7t} + te^{-t}\right)u(t)}$

– $F_{17}(p) = \dfrac{p - 1 + e^{-p}}{p^2\left(1 - e^{-p}\right)}$:

We isolate the periodicity term: $\dfrac{1}{1 - e^{-p}}$

Function $f_{17}(t)$ is periodic with a period of $T = 1$. The original over a single period is the reverse transform of $F_0(p) = \dfrac{p - 1 + e^{-p}}{p^2}$

This function is written as follows:

$f_0(t) = u(t) - t\, u(t) + (t - 1)u(t - 1)$

We make sure that this function is null outside of interval $t \in [0,1]$

Which gives us the solution: $\boxed{f_{17}(t) = \sum_{n=0}^{n=\infty} f_0(t - n)}$

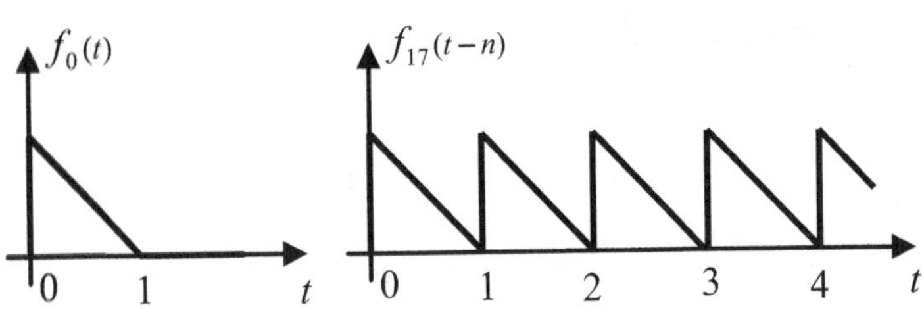

3.3.5. *Exercise 3.5*

We use the initial and final values theorem.

$$\boxed{\begin{array}{l} f(0) = \lim\limits_{p\to\infty}\left(pF(p)\right) = 3 \\ f(\infty) = \lim\limits_{p\to 0}\left(pF(p)\right) = -1 \end{array}}$$

3.3.6. *Exercise 3.6*

– $f_1(t) = e^{-\alpha t} u(t)$:

We refer to the tables to obtain:

$$F_1(p) = \frac{1}{p+\alpha}$$

And in the same way we also obtain:

$$F_2(p) = \frac{\alpha}{p(p+\alpha)}$$

$$F_3(p) = \frac{1}{(p+\alpha)^2}$$

– Initial and final values of these functions:

$$f_1(0) = \lim_{p\to\infty}\left(p\frac{1}{p+\alpha}\right) = 1$$

$$f_1(\infty) = \lim_{p\to 0}\left(p\frac{1}{p+\alpha}\right) = 0$$

In the same way, we also obtain:

$$f_2(0) = \lim_{p\to\infty}\left(p\frac{\alpha}{p(p+\alpha)}\right) = 0$$

$$f_2(\infty) = \lim_{p\to 0}\left(p\frac{\alpha}{p(p+\alpha)}\right) = 1$$

$$f_3(0) = \lim_{p\to\infty}\left(p\frac{1}{(p+\alpha)^2}\right) = 0$$

$$f_1(\infty) = \lim_{p\to 0}\left(p\frac{1}{(p+\alpha)^2}\right) = 0$$

– Area under the curve:

$$A(t)=\int_0^t f(x)dx \Rightarrow \mathcal{L}\{A(t)\}=\frac{F(p)}{p}$$

$A=\int_0^\infty f(x)dx = A(\infty)$, by applying the final value theorem:

$$\boxed{A=\lim_{p\to 0}\left(p\frac{F(p)}{p}\right)=\lim_{p\to 0}\left(F(p)\right)}$$

Hence: $\boxed{A_1=\frac{1}{\alpha} \qquad A_2=\infty \qquad A_3=\frac{1}{\alpha^2}}$

3.3.7. *Exercise 3.7*

– Calculation for $F_1(p)$:

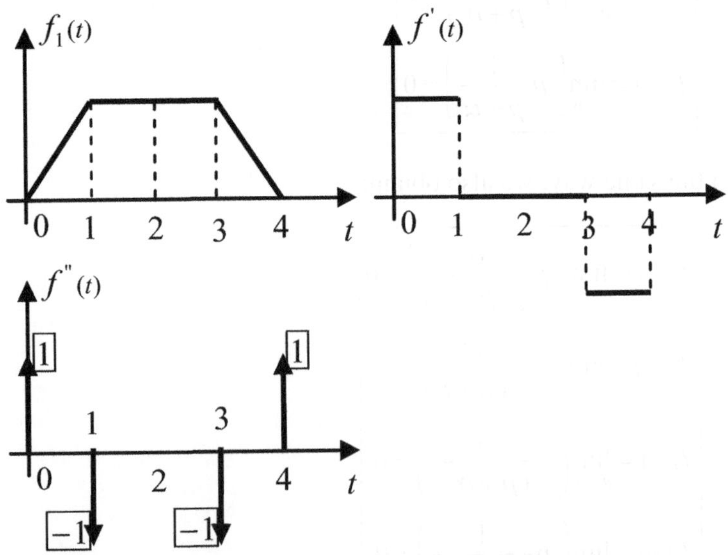

Function $f_1(t)$ can be posed as follows:

$$f_1(t) = t\,u(t) - (t-1)u(t-1) - (t-3)u(t-3) + (t-4)u(t-4)$$

Hence: $$\boxed{F_1(p) = \frac{1}{p^2}\left(1 - e^{-p} - e^{-3p} + e^{-4p}\right)}$$

We can also write:

$$f_1'(t) = u(t) - u(t-1) - u(t-3) + u(t-4)$$

Hence: $$\boxed{pF_1(p) = \frac{1}{p}\left(1 - e^{-p} - e^{-3p} + e^{-4p}\right)}$$

We can also write:

$$f_1''(t) = \delta(t) - \delta(t-1) - \delta(t-3) + \delta(t-4)$$

Hence: $$\boxed{p^2 F_1(p) = 1 - e^{-p} - e^{-3p} + e^{-4p}}$$

– Calculation of $F_2(p)$:

Using direct calculation:

$$F_2(p) = \int_0^1 t^2 e^{-pt}\,dt + \int_1^2 (2-t)e^{-pt}\,dt$$

Either integrating part by part:

$$\boxed{F_2(p) = \frac{2}{p^3}\left(1 - e^{-p}\right) - \frac{e^{-p}}{p^2}\left(3 - e^{-p}\right)}$$

– Calculation for $F_3(p)$:

Function $f_3(t)$ can be written:

$$f_3(t) = u(t) + u(t-1) + u(t-2) - u(t-3) - u(t-4) - u(t-5)$$

Hence: $\boxed{F_3(p)=\frac{1}{p}\left(1+e^{-p}+e^{-2p}-e^{-3p}-e^{-4p}-e^{-5p}\right)}$

We can also write:

$$f_3'(t)=\delta(t)+\delta(t-1)+\delta(t-2)-\delta(t-3)-\delta(t-4)-\delta(t-5)$$

Hence: $\boxed{pF_3(p)=1+e^{-p}+e^{-2p}-e^{-3p}-e^{-4p}-e^{-5p}}$

– Calculation for $F_4(p)$:

Function $f_4(t)$ is periodic with a period of $T=1$.

The function used to represent a single period is:

$$f_{04}(t)=\begin{cases}0 & t<0\\ e^{-2t} & t\in[0,1]\\ 0 & t>1\end{cases}$$

$$F_{04}(p)=\int_0^1 e^{-2t}e^{-pt}\,dt=\frac{1}{p+2}\left(1-e^{-(p+2)}\right)$$

$$f_4(t)=\sum_{n=0}^{n=\infty} f_{04}(t)\ \Rightarrow\ F_4(p)=\frac{F_{04}(p)}{1-e^{-p}}$$

Hence: $\boxed{F_4(p)=\frac{1}{p+2}\frac{1-e^{-(p+2)}}{1-e^{-p}}}$

– Calculation for $F_5(p)$:

Function $f_5(t)$ periodic with a period of $T=2$.

The function used to represent of a single period is:

$$f_{05}(t)=\sin 2\pi t\,u(t)-\sin 2\pi(t-1)\,u(t-1)$$

$$F_{05}(p)=\frac{2\pi}{p^2+4\pi^2}\left(1-e^{-p}\right)$$

Hence: $\boxed{F_5(p)=\frac{2\pi}{p^2+4\pi^2}\frac{1-e^{-p}}{1-e^{-2p}}}$

– Calculation for $F_6(p)$:

Function $f_6(t)$ is periodic with a period of $T=4$.

The function used to represent of a single period is:

$$f_{06}(t)=f(t)\sin 200\pi t$$

with $\quad f(t)=\frac{2}{3}t\,u(t)-\frac{4}{3}\left(t-\frac{3}{2}\right)u(t-\frac{3}{2})+\frac{2}{3}(t-3)u(t-3)$

$$\mathcal{L}\{f(t)\}=F(p)=\frac{2}{3p^2}\left(1-2e^{-\frac{3}{2}p}+e^{-3p}\right)=\frac{2}{3p^2}\left(1-e^{-\frac{3}{2}p}\right)^2$$

We know that: $\mathcal{L}\left\{f(t)e^{p_0t}\right\}=F(p-p_0)$

We can write: $f(t)\sin\omega_0 t=f(t)\frac{e^{j\omega_0t}-e^{-j\omega_0t}}{2j}$

Hence:

$$\mathcal{L}\{f_{06}(t)\}=\frac{1}{2j}\left(F(p-j\omega_0)-F(p+j\omega_0)\right)$$

$$=\frac{1}{2j}\left(\frac{2}{3(p-j\omega_0)^2}\left(1-e^{-\frac{3}{2}(p-j\omega_0)}\right)^2-\frac{2}{3(p+j\omega_0)^2}\left(1-e^{-\frac{3}{2}(p+j\omega_0)}\right)^2\right)$$

We note that $e^{-\frac{3}{2}(p\pm j200\pi)}=e^{-\frac{3}{2}p}$

$$F_{06}(p)=\frac{\left(1-e^{-\frac{3}{2}p}\right)^2}{3}\frac{400\pi}{p^2+(200\pi)^2}$$

Hence: $\boxed{F_5(p)=\frac{800\pi}{3}\frac{p}{\left(p^2+(200\pi)^2\right)^2}\frac{\left(1-e^{-\frac{3}{2}p}\right)^2}{1-e^{-4p}}}$

3.3.8. *Exercise 3.8*

– $y''+4y'+4y=2e^{-2x}$

We write:

$$\mathcal{L}\{y(x)\}=Y(p)$$
$$\mathcal{L}\{y'(x)\}=pY(p)-y(0)=pY(p)-2$$
$$\mathcal{L}\{y''(x)\}=p^2Y(p)-py(0)-y'(0)=p^2Y(p)-2p-5$$

In the same way: $\mathcal{L}\{e^{-2x}\}=\frac{1}{p+2}$

The Laplace transform of the differential equation appears as follows:

$$p^2Y-2p-5+4pY-8+4Y=\frac{2}{p+2}$$

Hence: $Y=\frac{2}{(p+2)^3}+\frac{2p}{(p+2)^2}+\frac{13}{(p+2)^2}$

Referring to the tables, we find:

$$\mathcal{L}\left\{x^n e^{-2x} u(x)\right\} = \frac{n!}{(p+2)^{n+1}}$$

Hence: $\boxed{y(x) = \left(x^2 + 9x + 2\right)e^{-2x}\, u(x)}$

– We take the Laplace transforms of both members and obtain:

$$p^4Y + 4p^3Y + 8p^2Y + 8pY + 3Y = \frac{-1}{p+1}$$

Hence:

$$Y = \frac{-1}{(p+1)(p^4 + 4p^3 + 8p^2 + 8p + 3)} = \frac{-1}{(p+1)^3(p^2 + 2p + 3)}$$

Which we also put into the following form:

$$Y = \frac{K_0}{(p+1)^3} + \frac{K_1}{(p+1)^2} + \frac{K_2}{(p+1)} + \frac{Ap+B}{(p^2+2p+3)}$$

We obtain: $K_0 = -\frac{1}{2}$, $K_1 = 0$, $K_2 = \frac{1}{4}$, $A = B = -\frac{1}{4}$ and $\frac{p+1}{p^2+2p+3} = \frac{p+1}{(p+1)^2+2}$, the reverse transform of which is: $e^{-x}\cos\sqrt{2}x\,u(x)$

Hence: $\boxed{y(x) = \left(1 - x^2 - \cos\sqrt{2}x\right)\frac{e^{-x}}{4}u(x)}$

This case is similar to the previous one, we obtain:

– $\boxed{y(x) = \left(xe^{-x} - e^{-x} + e^{-2x}\right)u(x)}$

3.3.9. *Exercise 3.9*

– We take the Laplace transforms of both equations:

$$\begin{cases} p^2Y_1 + 2pY_2 + 2Y_1 - 2Y_2 = \dfrac{10}{p} \\ p^2Y_2 + 2pY_1 + 2Y_2 - 2Y_1 = 0 \end{cases}$$

We find ourselves once again faced with a two system linear equation with two unknowns which we choose to solve using the Cramer method.

We obtain:
$$\begin{cases} Y_1(p) = \dfrac{10(p^2+2)}{p^2(p+2)(p^2-2p+4)} \\ Y_2(p) = \dfrac{-20(p-1)}{p^2(p+2)(p^2-2p+4)} \end{cases}$$

After expansion into partial-fraction, we obtain:

$$\boxed{\begin{cases} y_1(x) = \dfrac{5}{4}\left(2x + e^{-2x} - e^{x}\left(\cos\sqrt{3}x - \dfrac{1}{\sqrt{3}}\sin\sqrt{3}x\right)\right)u(x) \\ y_2(x) = \dfrac{5}{4}\left(2x - 2 + e^{-2x} - e^{x}\left(\cos\sqrt{3}x - \dfrac{1}{\sqrt{3}}\sin\sqrt{3}x\right)\right)u(x) \end{cases}}$$

– We proceed the same way again and obtain:

$$\boxed{\begin{cases} y_1(x) = \dfrac{1}{2}\left(x - e^{-x}\sin x\right)u(x) - \dfrac{1}{2}\left(x - 2 - e^{-(x-2)}\sin(x-2)\right)u(x-2) \\ y_2(x) = \dfrac{1}{2}\left(x - 1 + e^{-x}\cos x\right)u(x) - \dfrac{1}{2}\left(x - 3 + e^{-(x-2)}\sin(x-2)\right)u(x-2) \end{cases}}$$

3.3.10. *Exercise 3.10*

We must determine the Laplace transform of the input signal $E(p)$ and the transfer function of circuit $H(p)$:

The input signal is periodic with a period of $T = 1$. The signal over a single period can be posed as follows:

$$e_0(t) = u(t) - u(t - \frac{1}{4})$$

Hence: $$\boxed{E(p) = \frac{1}{p}\frac{1-e^{-\frac{p}{4}}}{1-e^{-p}}}$$

The transfer function for the circuit is easy to obtain:

$$H(p) = \frac{2}{(p+1)(p+2)}$$

We then obtain the Laplace transform of the output signal:

$$\boxed{S(p) = H(p)E(p) = \frac{1}{p}\frac{2}{(p+1)(p+2)}\frac{1-e^{-\frac{p}{4}}}{1-e^{-p}}}$$

We then show the transient state and the steady state under the following form:

$$S(p) = \frac{C_1}{(p+1)} + \frac{C_2}{(p+2)} + \frac{\Phi(p)}{1-e^{-p}}$$

Factor $\frac{1}{1-e^{-p}}$ indicates that the steady state is periodic with a period of $T = 1$.

– Studying the transient state.

The characteristic coefficients of the transient state are worth:

$$C_1 = -2e^{-\frac{3}{8}} \frac{sh\frac{1}{8}}{sh\frac{1}{2}} \approx -0.33059 \qquad C_2 = e^{-\frac{3}{4}} \frac{sh\frac{1}{4}}{sh 1} \approx 0.10153$$

The Laplace transform of the transient state is therefore:

$$\boxed{S_T(p) = \frac{C_1}{(p+1)} + \frac{C_2}{(p+2)}}$$

The time expression of the transient state is written as follows:

$$\boxed{s_T(t) = \left(C_1 e^{-t} + C_2 e^{-2t}\right) u(t)}$$

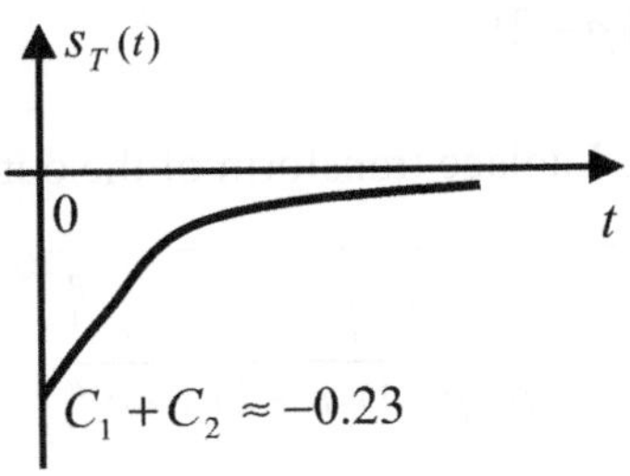

– Studying the steady state.

The characterisitic function of steady state is easy to obtain:

$$\Phi(p) = \frac{2\left(1 - e^{-\frac{p}{4}}\right)}{p(p+1)(p+2)} - \left(\frac{C_1}{p+1} + \frac{C_2}{p+2}\right)\left(1 - e^{-p}\right)$$

If we arrange this:

$$\Phi(p) = \left(\frac{1}{p} - \frac{2}{(p+1)} + \frac{1}{(p+2)}\right)\left(1 - e^{-\frac{p}{4}}\right) - \left(\frac{C_1}{p+1} + \frac{C_2}{p+2}\right)\left(1 - e^{-p}\right)$$

We write this under the following form:

$$\Phi(p) = \left(\frac{1}{p} - \frac{2+C_1}{(p+1)} + \frac{1-C_2}{(p+2)}\right) - \left(\frac{1}{p} - \frac{2}{(p+1)} + \frac{1}{(p+2)}\right)e^{-\frac{p}{4}} + \left(\frac{C_1}{p+1} + \frac{C_2}{p+2}\right)e^{-p}$$

From this we can establish the time-based expression of the steady state:

$$\varphi(t) = \left(1 - (2+C_1)e^{-t} + (1-C_2)e^{-2t}\right)u(t) - \left(1 - 2e^{-\left(t-\frac{1}{4}\right)} + e^{-2\left(t-\frac{1}{4}\right)}\right)u(t-\frac{1}{4})$$
$$+ \left(C_1 e^{-(t-1)} + C_2 e^{-2(t-1)}\right)u(t-1)$$

We study this function in three intervals:

$$t \in \left[0, \frac{1}{4}\right]$$

$$\varphi(t) = 1 - (2+C_1)e^{-t} + (1-C_2)e^{-2t} = 1 - 1.67e^{-t} + 0.9e^{-2t}$$

$$t \in \left[\frac{1}{4}, 1\right]$$

$$\varphi(t) = 1 - 1.67e^{-t} + 0.9e^{-2t} - 1 + 2e^{-\left(t-\frac{1}{4}\right)} - e^{-2\left(t-\frac{1}{4}\right)}$$

$$\varphi(t) = 0.89e^{-t} - 0.75e^{-2t}$$

$$t \in \left[\frac{1}{4}, \infty\right]$$

$$\varphi(t) = 0.89e^{-t} - 0.75e^{-2t} + C_1 e^{-(t-1)} + C_2 e^{-2(t-1)}$$

$$\varphi(t) = 0$$

This final result makes sense as the periodic element must be null outside of the period.

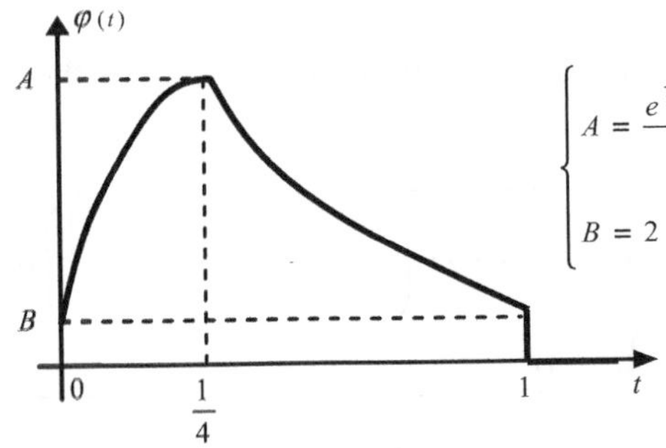

$$\begin{cases} A = \dfrac{e^{\frac{1}{2}} - e^{\frac{1}{4}}}{e-1} \approx 4.77 \\ B = 2\dfrac{e^{\frac{1}{4}} - 1}{e-1} - \dfrac{e^{\frac{1}{2}} - 1}{e^2 - 1} \approx 0.23 \end{cases}$$

The steady state which is periodic with a period of $T = 1$ corresponds to function $\varphi(t)$ that is identically repeated once every second.

$$\boxed{s_P(t) = \sum_{n=0}^{n=\infty} \varphi(t-n)}$$

The complete response to the circuit is written as follows:

$$\boxed{s(t) = s_T(t) + s_P(t) = s_T(t) + \sum_{n=0}^{n=\infty} \varphi(t-n)}$$

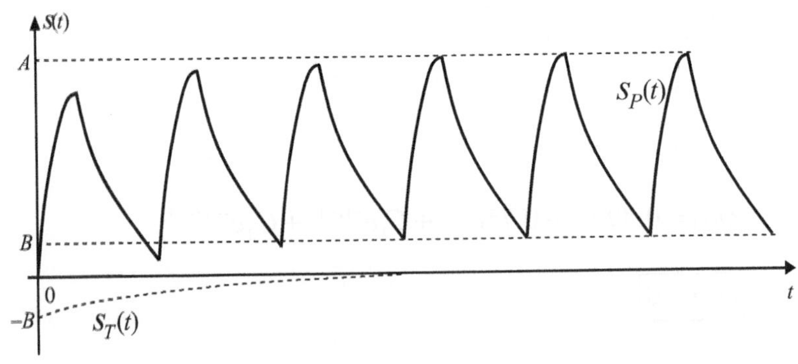

4

Integrals and Convolution Product

4.1. Theoretical background

4.1.1. *Analyzing linear systems using convolution integrals*

Consider the linear system of a transfer function $H(p)$. The Laplace transforms of input and output signals, $e(t)$ and $s(t)$, are written $E(p)$ and $S(p)$. These values are linked by:

$$\boxed{S(p) = H(p)E(p)} \qquad [4.1]$$

The reverse transform of $H(p)$, written $h(t)$, represents the *impulse response* of the system.

If the input into the system is a Dirac pulse, we can write:

$e(t) = \delta(t)$, then $E(p) = 1$ and $S(p) = H(p)$ hence $s(t) = h(t)$

In most cases, the response to the signal to a basic input signal is the reverse transform of $H(p)E(p)$, or

$$\boxed{s(t) = h(t) * e(t) = \int_{-\infty}^{+\infty} h(\tau)e(t-\tau)d\tau} \qquad [4.2]$$

$h(t) * e(t)$ is the convolution product defined by the convolution integral $\int_{-\infty}^{+\infty} h(\tau)e(t-\tau)d\tau$

The impulse response is sometimes called Green's function.

The poles of the convolution integral can be modified depending on the causality of $e(t)$ and $h(t)$.

$\int_{-\infty}^{+\infty} h(\tau)e(t-\tau)d\tau$	$e(t)$ and $h(t)$ any support
$\int_{-\infty}^{t} h(\tau)e(t-\tau)d\tau$	$e(t)$ causal $h(t)$ any support
$\int_{0}^{t} h(\tau)e(t-\tau)d\tau$	$e(t)$ and $h(t)$ causal

4.1.2. Convolution properties

4.1.2.1. *Commutativity*

$$f_1(t) * f_2(t) = f_2(t) * f_1(t) \quad [4.3]$$

4.1.2.2. *Distributivity*

$$f_1(t) * (f_2(t) + f_3(t)) = f_1(t) * f_2(t) + f_2(t) * f_3(t) \quad [4.4]$$

4.1.2.3. *Associativity*

$$f_1(t) * (f_2(t) * f_3(t)) = (f_1(t) * f_2(t)) * f_3(t) \quad [4.5]$$

4.1.2.4. *Derivative of convolution*

$$\frac{d}{dt}(f_1(t) * f_2(t)) = \frac{df_1(t)}{dt} * f_2(t) = f_1(t) * \frac{df_2(t)}{dt} \quad [4.6]$$

4.1.2.5. *Time-domain translation*

Let us set $p(t) = f_1(t) * f_2(t)$ the convolution product of two functions.

$$p(t-h) = f_1(t-h) * f_2(t) = f_1(t) * f_2(t-h) \quad [4.7]$$

4.1.2.6. *Fourier transform of a convolution product*

Consider $f_1(t) \xleftrightarrow{TF} F_1(j\omega)$ and $f_2(t) \xleftrightarrow{TF} F_2(j\omega)$

$$f_1(t) * f_2(t) \xleftrightarrow{TF} F_1(j\omega) F_2(j\omega) \quad [4.8]$$

4.1.3. *Graphical interpretation of the convolution product*

Let us consider two functions $f_1(t)$ and $f_2(t)$. The independent variable of the integration is τ. We can then outline $f_1(\tau)$ and $f_2(\tau)$, we obtain $f_2(-\tau)$ through symmetry of $f_2(\tau)$ against the ordinate axis.

$f_2(t-\tau)$ represents $f_2(-\tau)$ translated of amount t on the abscissa axis.

The value of the convolution is equal to the surface between the abscissa and $f_1(\tau) f_2(t-\tau)$.

Figure 4.1 shows an example outline. The hatched areas represent the value of the convolution product at times t_1 and t_2.

Figure 4.2 gives us a representation of the convolution product of two functions.

4.1.4. *Convolution of a function using a unit impulse*

Let us consider a function $f(t)$. The convolution using a Dirac impulse would appear as follows:

$$f(t) * \delta(t) = \int_{-\infty}^{+\infty} \delta(\tau) f(t-\tau) d\tau = f(t-\tau)\Big|_{\tau=0} = f(t)$$

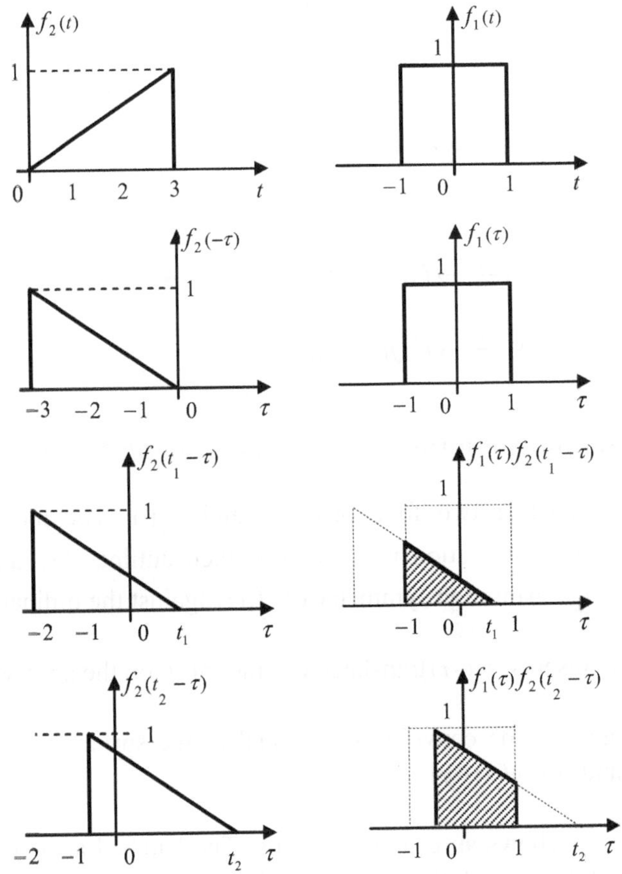

Figure 4.1. *Graphical representation of a convolution product*

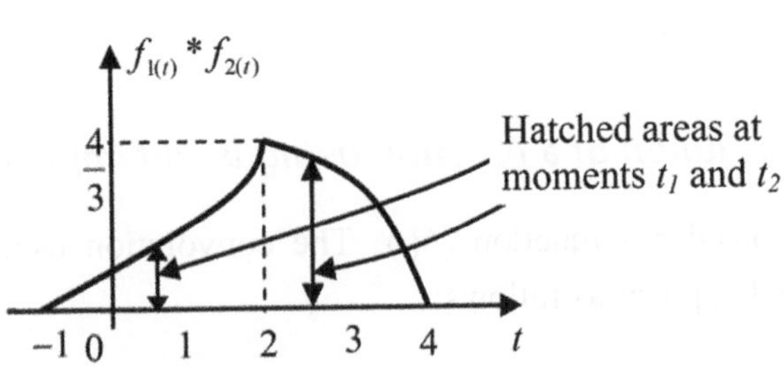

Figure 4.2. *Outline of the convolution product*

The convolution product of a function by a Dirac impulse is equal to said function:

$$\boxed{f(t) * \delta(t) = f(t)} \qquad [4.9]$$

From this, we establish the following:

$$\boxed{\begin{array}{l} f(t) * k\delta(t) = kf(t) \\ f(t) * \delta(t - t_0) = f(t - t_0) \\ f(t - t_1) * \delta(t - t_2) = f(t - (t_1 + t_2)) \end{array}} \qquad [4.10]$$

4.1.5. *Step response from a system*

A step response refers to the response to an input signal such as unit step $u(t)$.

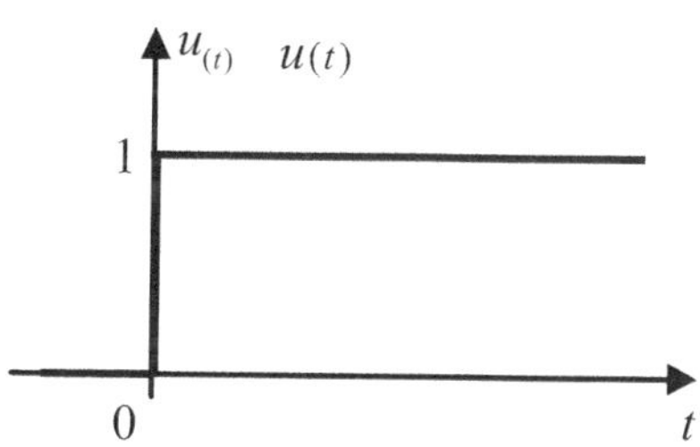

Figure 4.3. *Unit step*

We note $r(t)$ this response:

$$r(t) = h(t) * u(t)$$

The derivative of this response appears as follows:

$$\frac{dr(t)}{dt} = h(t) * \frac{du(t)}{dt} = h(t) * \delta(t)$$

The impulse response of a system is the derivative of the step response.

$$\begin{cases} h(t) = \dfrac{dr(t)}{dt} \\ r(t) = \displaystyle\int_0^t h(\tau)d\tau \end{cases} \qquad [4.11]$$

4.1.6. *Eigenfunction of a convolution operator*

An eigenfunction of a convolution operator is a function $x(t)$ such that:

$$x(t) * x(t) = kx(t)$$

From this we establish that the Fourier transform of $x(t)$ must be null or equal to a constant or a linear combination of both.

Let us consider the example represented in Figure 4.4:

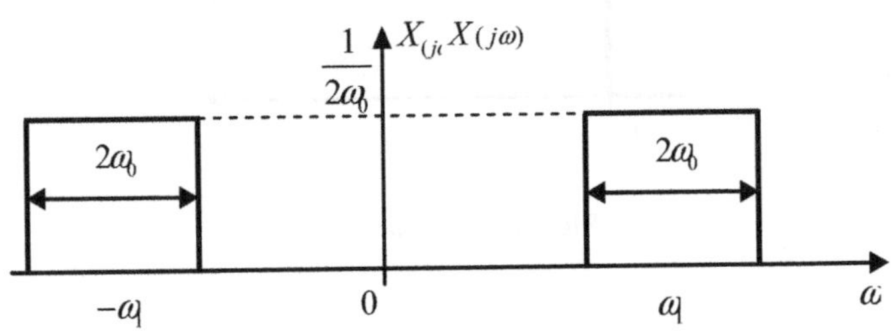

Figure 4.4. *Fourier transform of an eigenfunction*

The reverse transform of this function is $x(t)$:

$$x(t) = \frac{\sin \omega_0 t}{\omega_0 t} \cos \omega_1 t$$

If $\omega_0 = 0 \Rightarrow x(t) = \cos \omega_1 t$

The cosine function is thus an eigenfunction of the convolution operator.

4.2. Exercises

The exercises presented in this chapter can be found in B.P. Lathi [LAT 66].

4.2.1. *Exercise 4.1*

Determine the convolutions of $f_1(t) * f_2(t)$ for the following functions:

a)

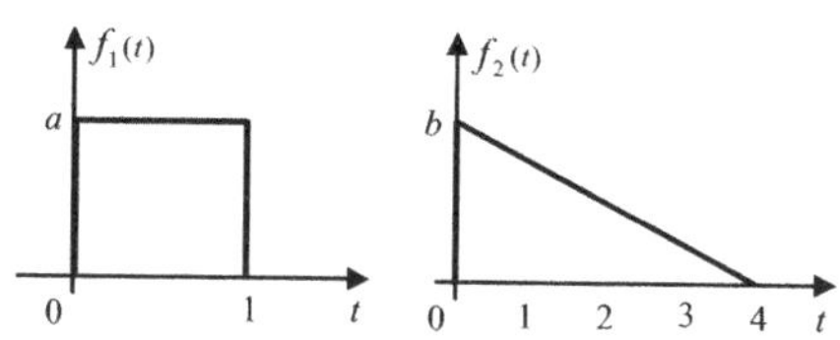

b)

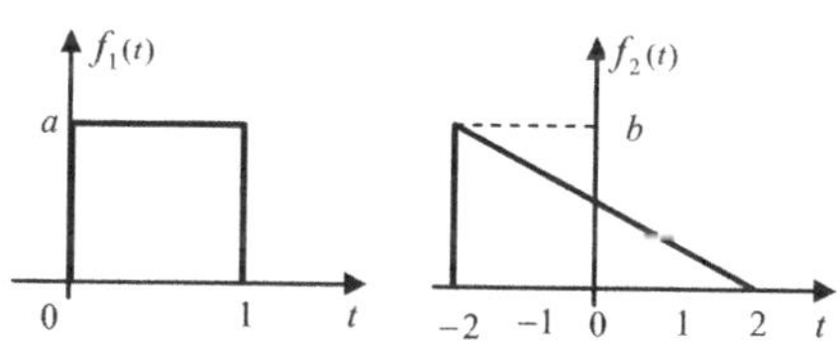

c)

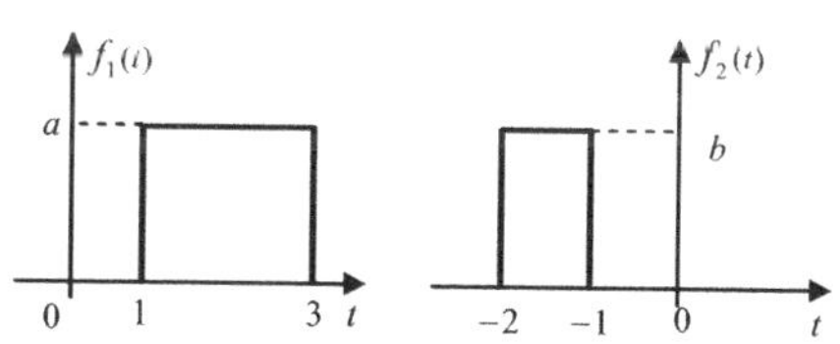

d)

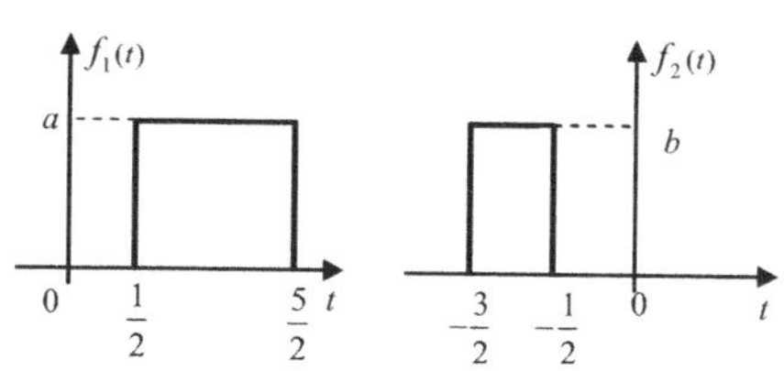

4.2.2. Exercise 4.2

Consider $h(t)$ the impulse response to a linear system and $e(t)$ the input signal. Determine the output signal using the convolution in the following cases:

a) $e(t) = tu(t)$ $\quad h(t) = u(t)$

b) $e(t) = tu(t)$ $\quad h(t) = e^{-at}u(t)$

c) $e(t) = e^{-at}u(t)$ $\quad h(t) = \left(1 - e^{-at}\right)u(t)$

d) $e(t) = e^{-a|t|}$ $\quad h(t) = u(t)$

e) $e(t) = e^{-a|t|}$ $\quad h(t) = e^{-at}u(t)$

f) $e(t) = te^{-b|t|}u(t)$ $\quad h(t) = te^{-at}u(t)$

g) $e(t)$ $\quad h(t) = e^{-t}u(t)$

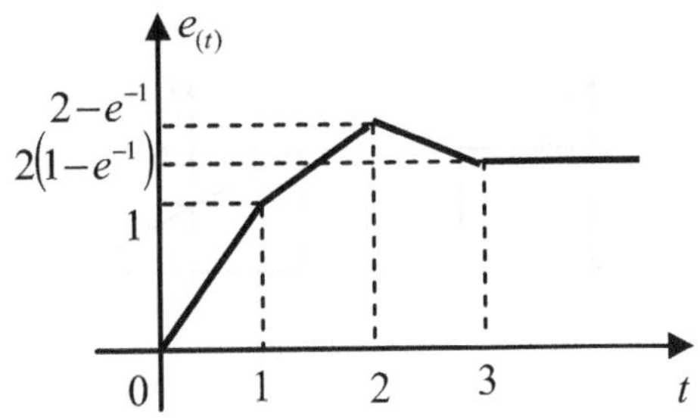

4.2.3. Exercise 4.3

Let us consider the product of two functions $f(t) = f_1(t)f_2(t)$. Determine the Fourier transform of $f(t)$ using the convolution in the frequency domain.

a) $f_1(t) = \cos\omega_1 t$ $\quad f_2(t) = \cos\omega_2 t$

b) $f_1(t) = \sum_{k=-\infty}^{k=+\infty} \delta(t - kT)$ $\quad f_2(t) = \dfrac{\sin\omega_C t}{\omega_C t}$ $\quad \dfrac{2\pi}{T} > 2\omega_C$

c) $f_1(t) = f_2(t) = \dfrac{\sin at}{\pi t}$

d) $f_1(t) = \left(\dfrac{\sin at}{\pi t}\right)^2 \quad f_2(t) = \dfrac{\sin at}{\pi t}$

4.2.4. *Exercise 4.4*

Consider a causal function $f_1(t)$, null if $t > t_0$.

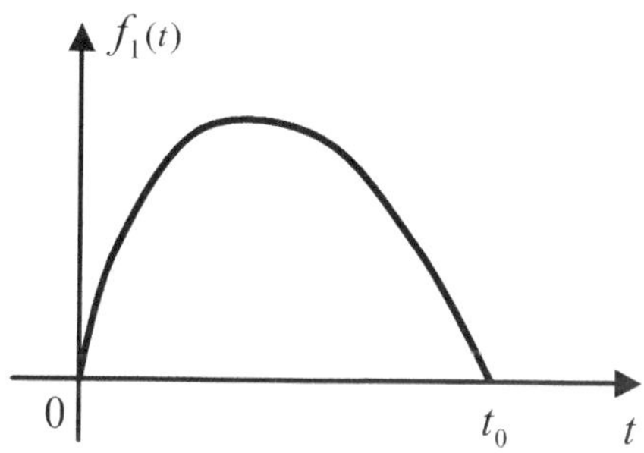

Consider also a function $f_2(t) = \sum_{k=-\infty}^{k=+\infty} \delta(t - kT)$ with $T > t_0$

Determine the convolution product of these two functions using the graph method and check the result using frequency analaysis.

4.2.5. *Exercise 4.5*

Consider any function $f(t)$. Determine function $\phi(t)$ such that the following equations can all be verified:

a) $f(t) = \dfrac{d}{dt}\left(f(t) * \phi(t)\right)$

b) $f(t) = \int_0^t \left(f(x) * \phi(x)\right) dx$

4.2.6. *Exercise 4.6*

Consider the following setup:

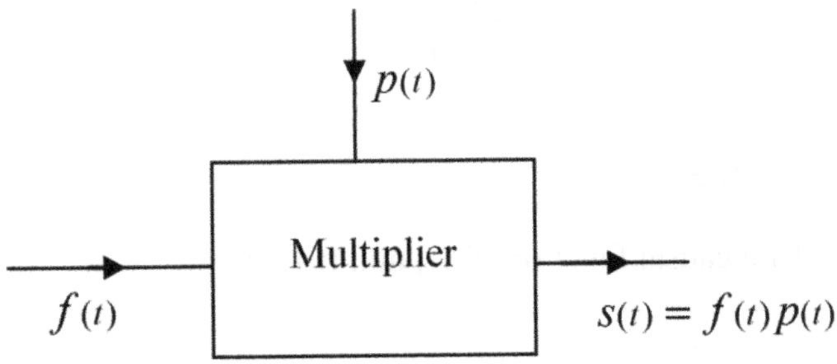

And the time and frequency representations of $f(t)$:

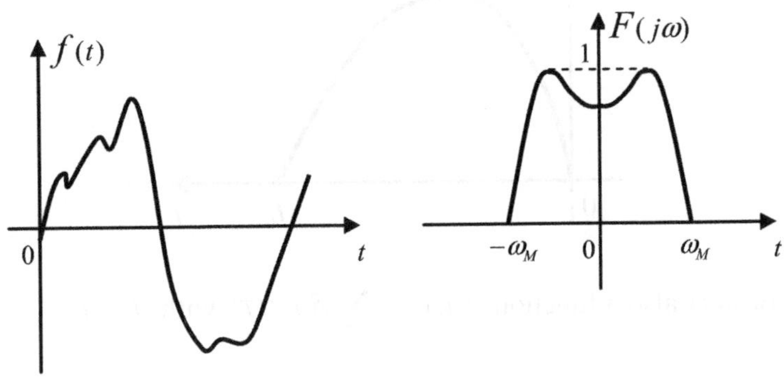

What happens to the spectrum of $s(t)$ when $p(t)$ takes on the following forms:

a) $p(t) = \cos \omega_0 t$

b) $p(t) = \sum_{n=-\infty}^{n=+\infty} C r_{\frac{T}{2}}(t - nT) \qquad T = \frac{2\pi}{\omega_0}$

c) $p(t)$ is any periodic function with a period of T

We suppose that $\omega_0 > 2\omega_M$.

4.3. Solutions to the exercises

4.3.1. *Exercise 4.1*

a) We use the graph method by tracing $f_1(\tau)$ and $f_2(t-\tau)$

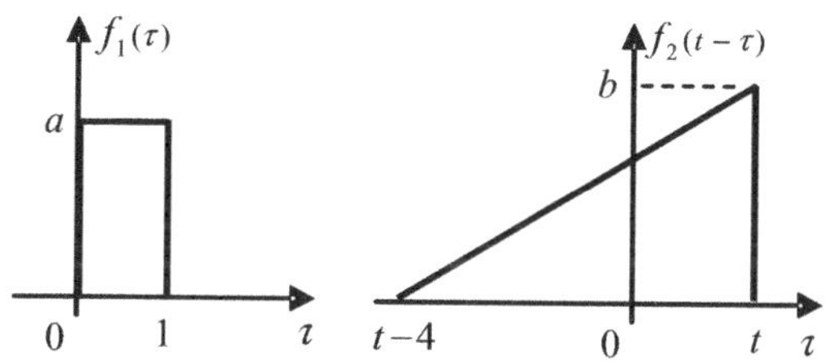

We determine the area inside the product curve $f_1(\tau)f_2(t-\tau)$ by varying τ by $-\infty$ and $+\infty$

– $t<0$

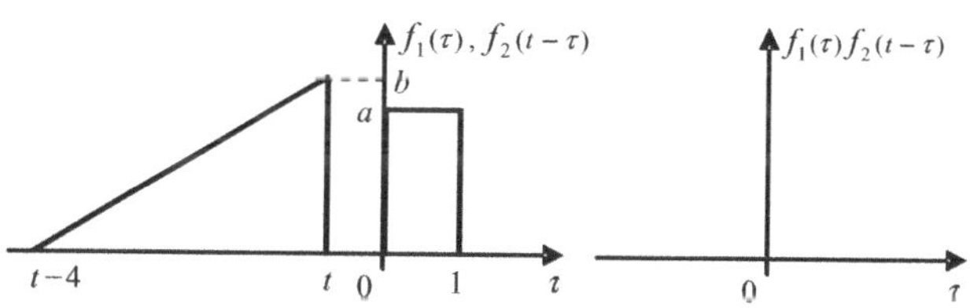

$$f_1(t)*f_2(t)=0$$

– $0<t<1$

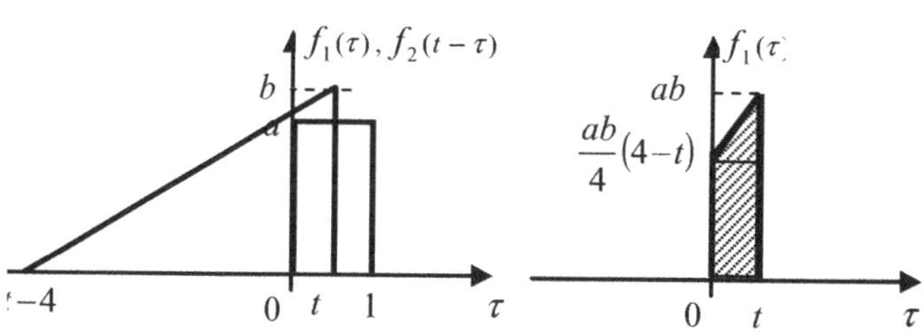

$$f_1(t)*f_2(t)=abt\left(1-\frac{t}{8}\right)$$

– $1 < t < 4$

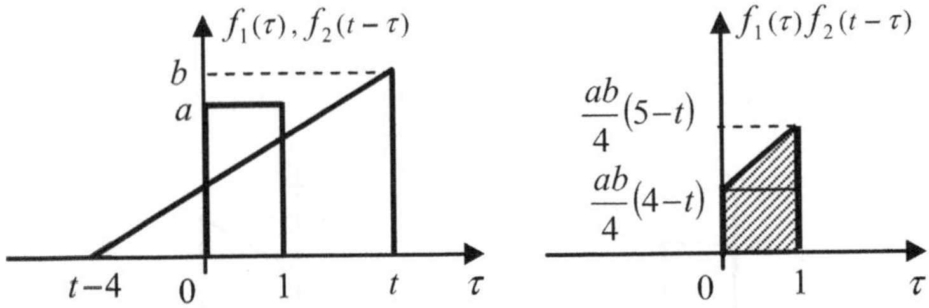

$$f_1(t) * f_2(t) = \frac{ab}{8}(9 - 2t)$$

– $4 < t < 5$

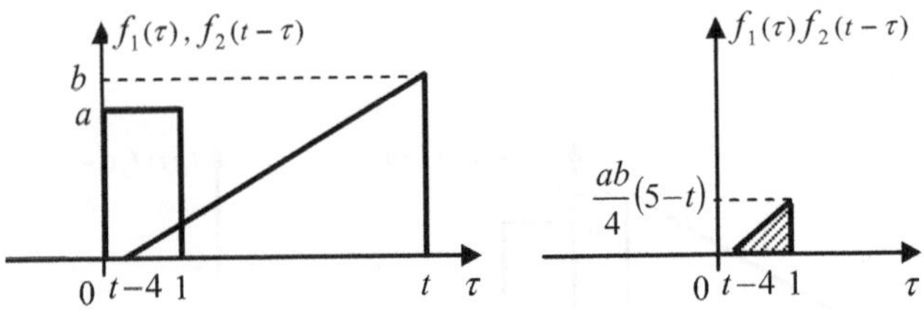

$$f_1(t) * f_2(t) = \frac{ab}{8}(5 - t)^2$$

– $t > 5$

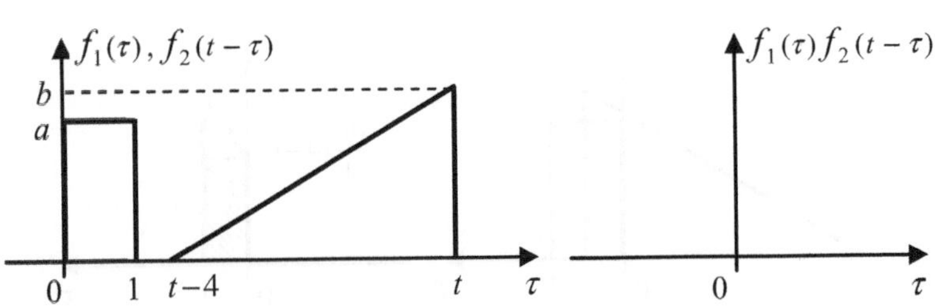

$$f_1(t) * f_2(t) = 0$$

We can then trace $f_1(t) * f_2(t)$

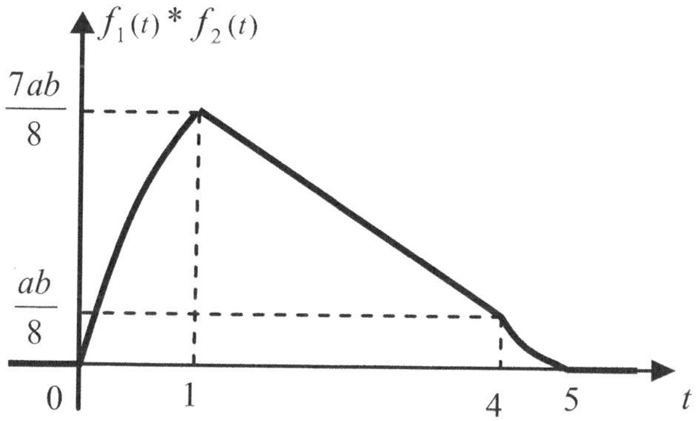

b) In this case, $f_2(t)$ corresponds to the one in question *a* translated of 2 to the left, so the convolution is the same.

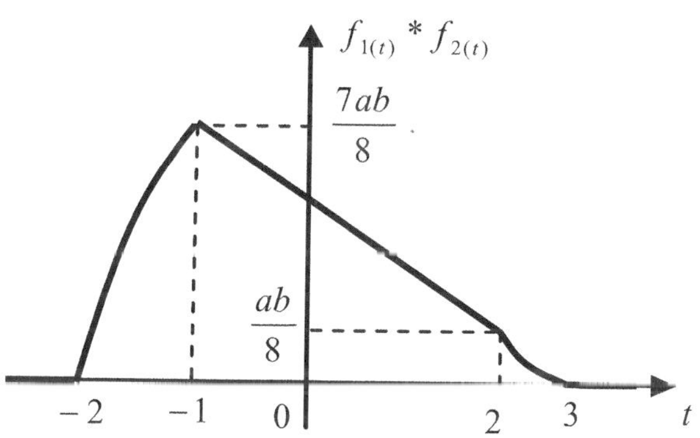

c) We will only show the results with p the convolution product.

$t < -1$	$p = 0$
$-1 < t < 0$	$p = ab(t+1)$
$0 < t < 1$	$p = ab$
$1 < t < 2$	$p = ab(2-t)$
$t > 2$	$p = 0$

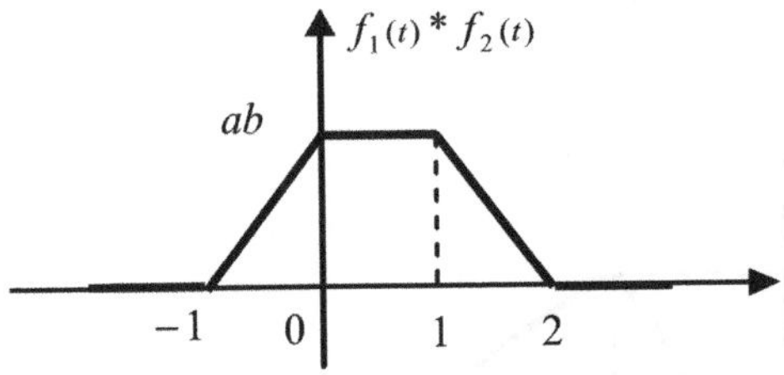

d) For question c, we notice a translation of $-\frac{1}{2}$ for $f_1(t)$ and f

of $+\frac{1}{2}$ for $f_2(t)$.

The convolution is still the same.

4.3.2. *Exercise 4.2*

The output is expressed as a convolution:

$$s(t) = h(t) * e(t) = \int_{-\infty}^{+\infty} h(t-\tau)e(\tau)d\tau$$

a) Let us first use an analytical calculation.

$$e(t) = tu(t) \quad h(t) = u(t) \quad s(t) = \int_{-\infty}^{+\infty} \tau u(\tau)u(t-\tau)d\tau$$

$$\begin{array}{ll} u(\tau) = 0 & \tau < 0 \\ u(\tau) = 1 & \tau > 0 \\ u(t-\tau) = 0 & \tau > t \\ u(t-\tau) = 1 & \tau < t \end{array}$$

Hence: $\boxed{s(t) = \int_0^t \tau d\tau = \frac{t^2}{2}u(t)}$

By using the graph method.

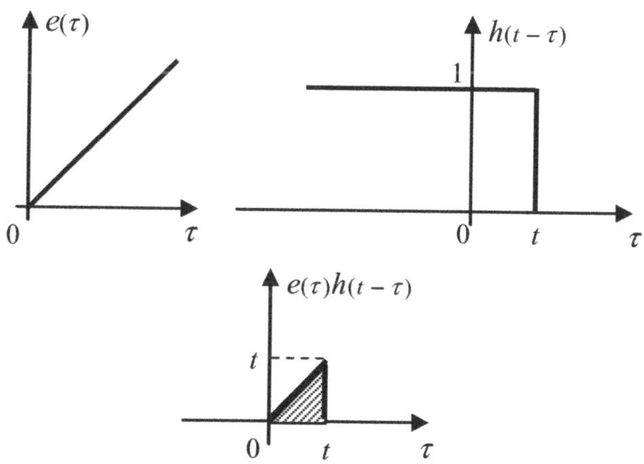

The hatched area is equal to $\frac{t^2}{2}u(t)$.

For the following questions, we will only show the results.

b)

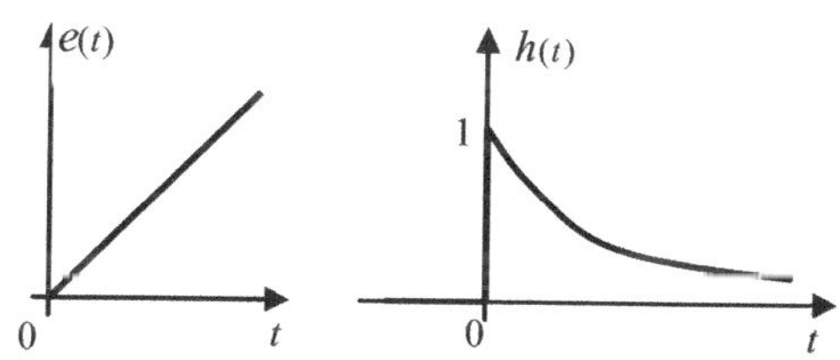

$$s(t) = \frac{1}{a^2}\left(-1 + at + e^{-at}\right) u(t)$$

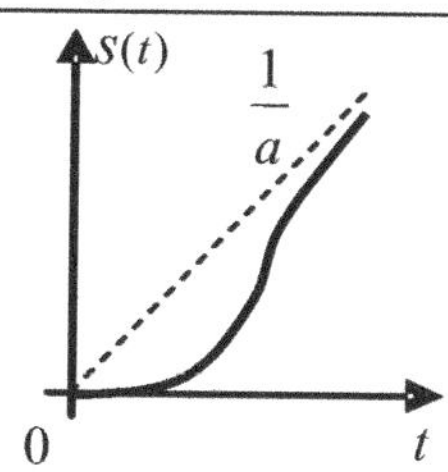

c)

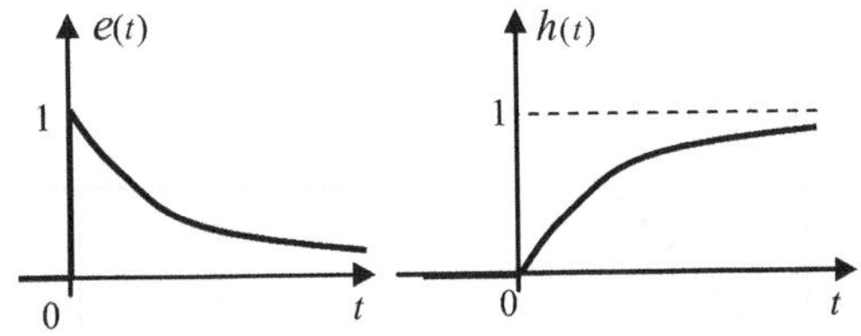

$$s(t) = \frac{1}{a}\left(1 - e^{-at} - ate^{-at}\right)u(t)$$

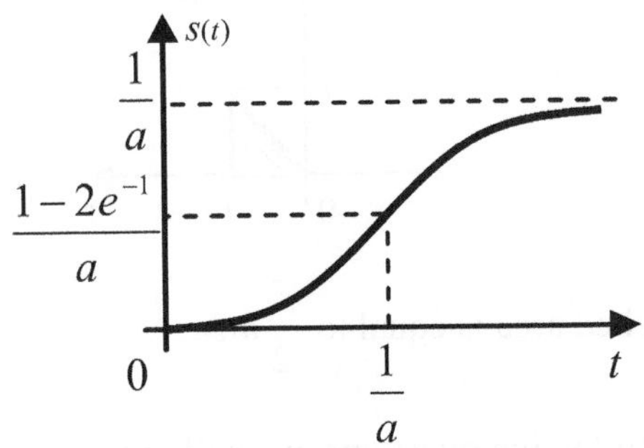

d)

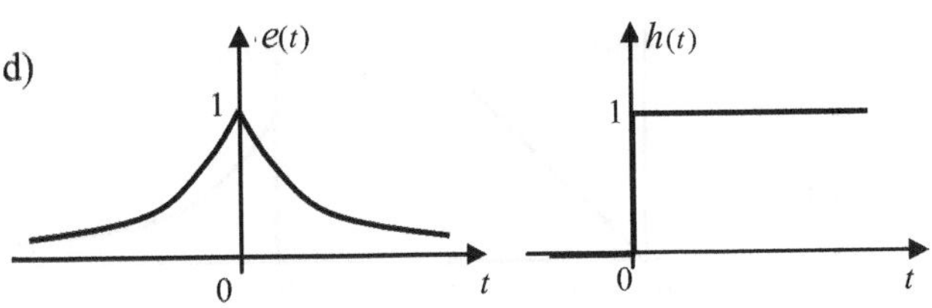

$$s(t) = \frac{1}{a} + \frac{1}{a}\left(1 - e^{-a|t|}\right)sign(t)$$

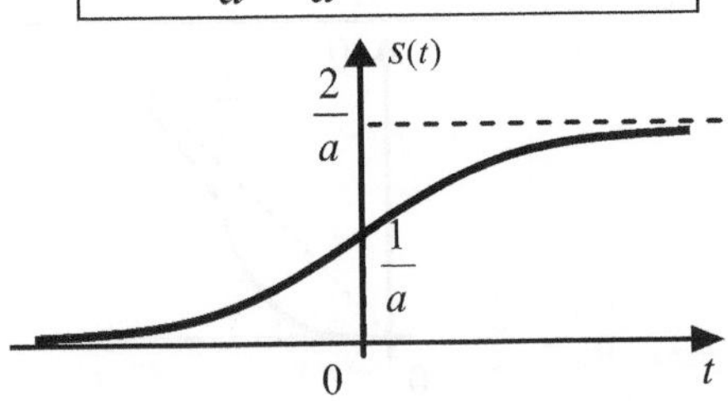

e)

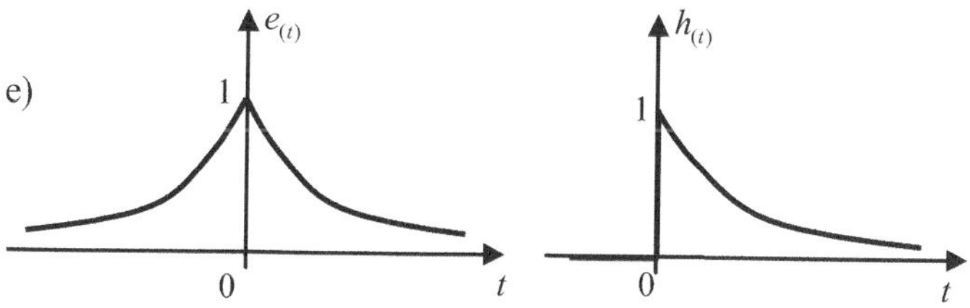

$$s(t) = \begin{cases} \dfrac{e^{at}}{2a} & t < 0 \\ \left(\dfrac{1}{2a} + t\right)e^{-at} & t > 0 \end{cases}$$

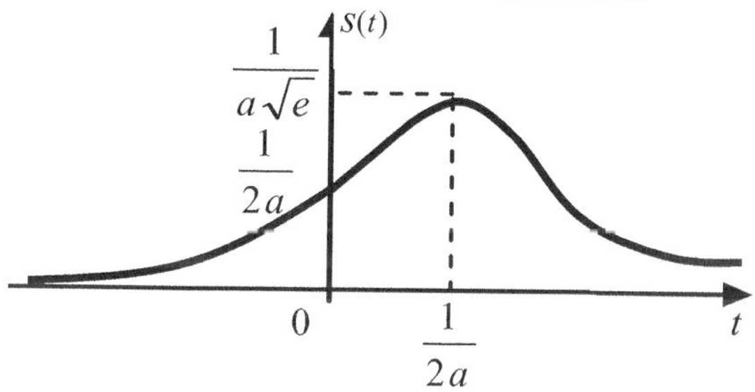

f)

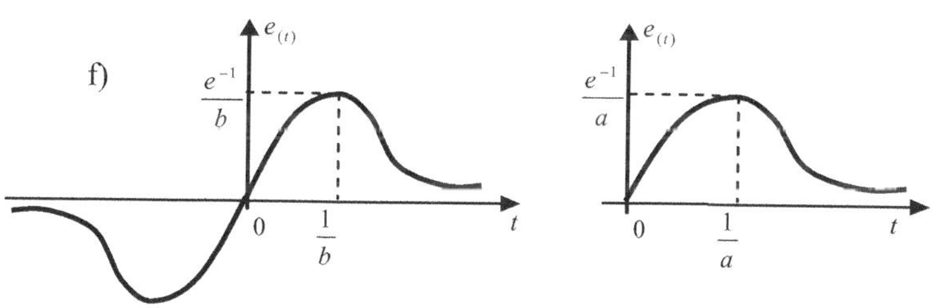

$$s(t) = \begin{cases} \dfrac{e^{bt}}{(a+b)^2}\left(t - \dfrac{2}{a+b}\right) & t < 0 \\ \dfrac{t}{(a-b)^2}\left(e^{-at} + e^{-bt}\right) + \dfrac{2}{(a-b)^3}\left(e^{-at} - e^{-bt}\right) - \dfrac{e^{-at}}{(a+b)^2}\left(t + \dfrac{2}{a+b}\right) & t > 0 \end{cases}$$

g)
$$\begin{array}{|ll|}\hline t<0 & s(t)=0 \\ 0<t<1 & s(t)=t-1-e^{-t} \\ 1<t<2 & s(t)=\left(1-e^{-1}\right)(t-1)-e^{-1} \\ 2<t<3 & s(t)=2\left(1+e^{-1}\right)-e^{-1}t-e^{2}e^{-t} \\ t>3 & s(t)=2\left(1-e^{-1}\right) \\ \hline \end{array}$$

4.3.3. *Exercise 4.3*

If $f(t)=f_1(t)f_2(t)$ then $F(j\omega)=\dfrac{1}{2\pi}\left(F_1(j\omega)*F_2(j\omega)\right)$

a) $\begin{cases} f_1(t)=\cos\omega_1 t \xleftrightarrow{TF} F_1(j\omega)=\pi\left(\delta(\omega-\omega_1)+\delta(\omega+\omega_1)\right) \\ f_2(t)=\cos\omega_2 t \xleftrightarrow{TF} F_2(j\omega)=\pi\left(\delta(\omega-\omega_2)+\delta(\omega+\omega_2)\right) \end{cases}$

Hence:

$$F(j\omega)=F_1(j\omega)*F_2(j\omega)$$
$$=\frac{\pi}{2}\left(\delta(\omega-\omega_1-\omega_2)+\delta(\omega-\omega_1+\omega_2)+\delta(\omega+\omega_1-\omega_2)+\delta(\omega+\omega_1+\omega_2)\right)$$

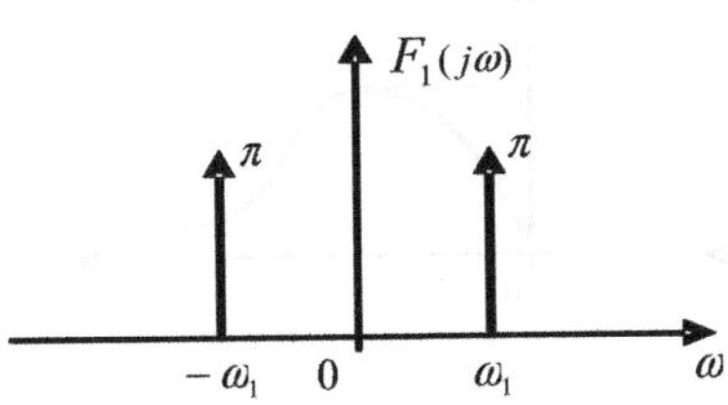

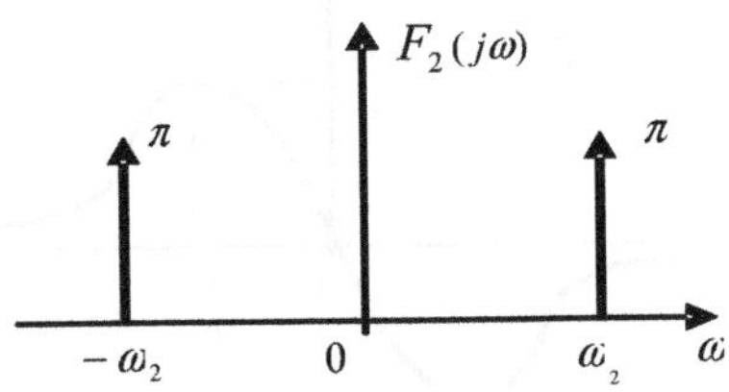

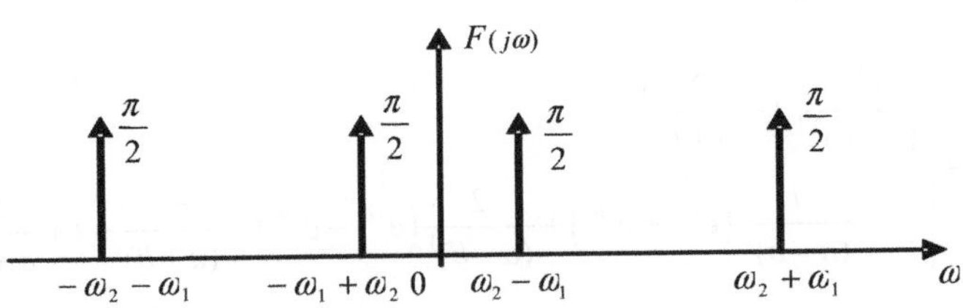

b) $$\begin{cases} f_1(t) = \displaystyle\sum_{n=-\infty}^{n=+\infty} \delta(t-kT) \xleftrightarrow{TF} F_1(j\omega) = \frac{2\pi}{T} \sum_{n=-\infty}^{n=+\infty} \delta(\omega - k\frac{2\pi}{T}) \\ f_2(t) = \dfrac{\sin \omega_C t}{\omega_C t} \xleftrightarrow{TF} F_2(j\omega) = \dfrac{\pi}{\omega_C} Cr_{2\omega_C}(\omega) \end{cases}$$

Hence: $$F(j\omega) = \frac{1}{2\pi}\left(\frac{\pi}{\omega_C} Cr_{2\omega_C}(\omega)\right) * \left(\frac{2\pi}{T}\sum_{n=-\infty}^{n=+\infty} \delta(\omega - k\frac{2\pi}{T})\right)$$

We can also set $\omega_0 = \dfrac{2\pi}{T}$

$$F(j\omega) = \frac{1}{2}\frac{\omega_0}{\omega_C}\sum_{n=-\infty}^{n=+\infty} Cr_{2\omega_C}(\omega - k\omega_0)$$

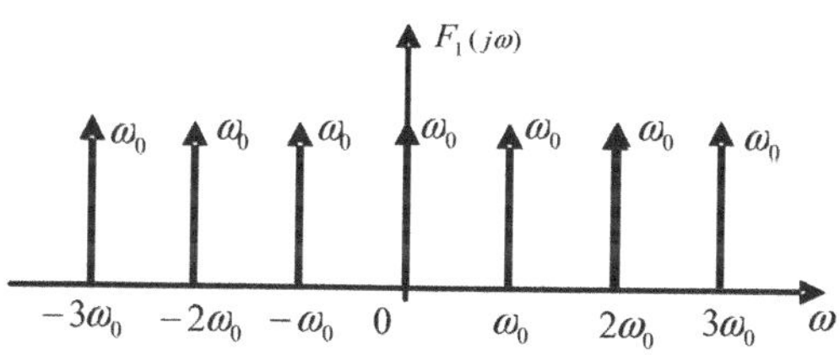

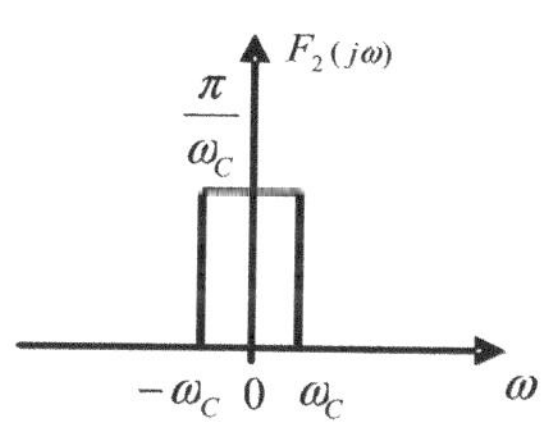

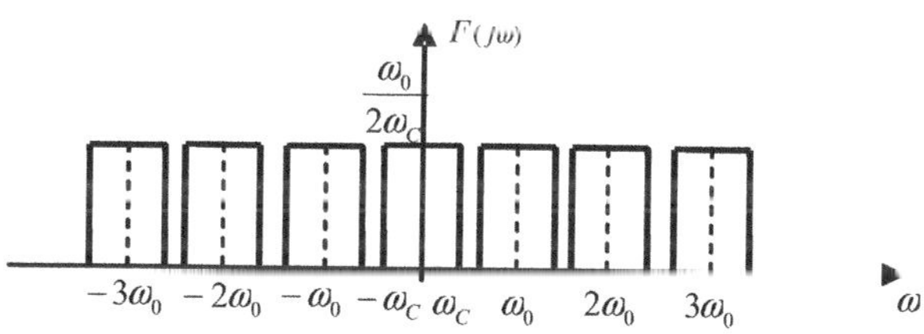

We will discuss in detail these types of exercises in Chapter 6 which is dedicated to sampling.

c) $$f_1(t) = f_2(t) = \frac{\sin at}{\pi t} \xleftrightarrow{TF} F_1(j\omega) = F_2(j\omega) = Cr_{2a}(\omega)$$

$$\text{Hence: } F(j\omega) = \begin{cases} \dfrac{1}{2\pi}\left(2a - |\omega|\right) & \omega \in [-2a, +2a] \\ 0 & \omega \notin [-2a, +2a] \end{cases}$$

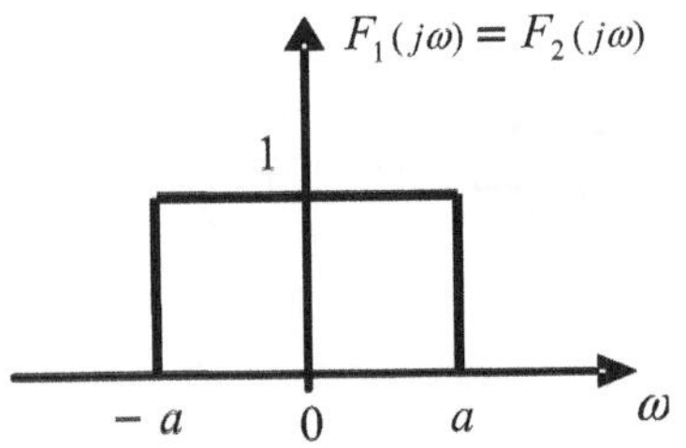

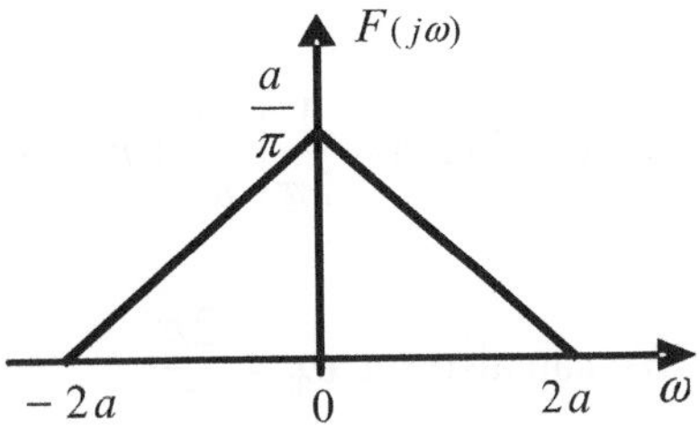

d) $$\begin{cases} f_1(t) = \left(\dfrac{\sin at}{\pi t}\right)^2 \xleftrightarrow{TF} F_1(j\omega) = \begin{cases} \dfrac{1}{2\pi}\left(2a - |\omega|\right) & \omega \in [-2a, +2a] \\ 0 & \omega \notin [-2a, +2a] \end{cases} \\ f_2(t) = \dfrac{\sin at}{\pi t} \xleftrightarrow{TF} F_2(j\omega) = Cr_{2a}(\omega) \end{cases}$$

$$\text{Hence: } F(j\omega) = \begin{cases} 0 & |\omega| \notin [3a, +\infty] \\ \dfrac{\left(3a - |\omega|\right)^2}{8\pi^2} & |\omega| \in [a, 3a] \\ \dfrac{3a^2 - \omega^2}{4\pi^2} & |\omega| \in [0, a] \end{cases}$$

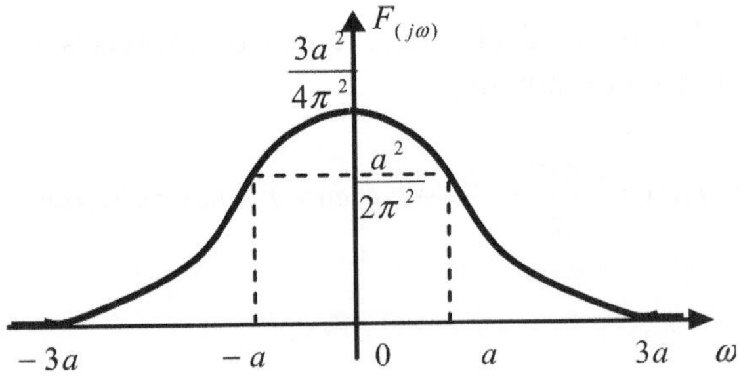

4.3.4. *Exercise 4.4*

Let us plot $f_2(\tau)$ and $f_1(t-\tau)$

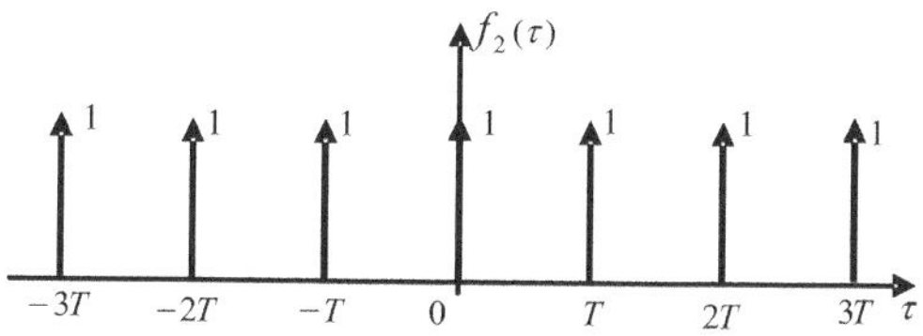

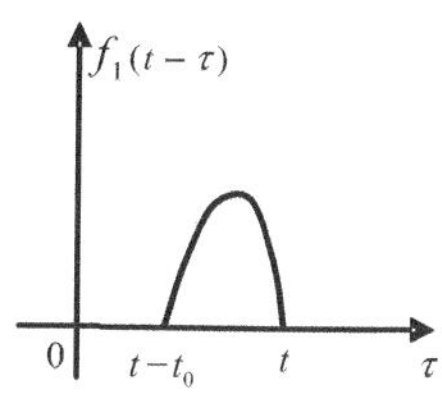

There are two possibilities:

– $t_0 < t < T$ modulo T

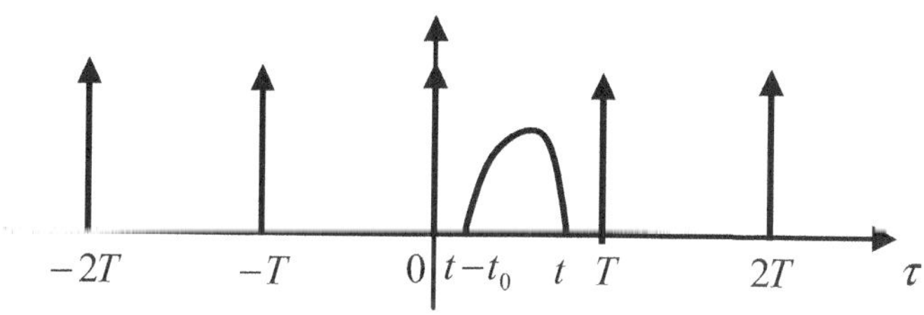

$$f_1(t) * f_2(t) = 0$$

– $t > T$ modulo T

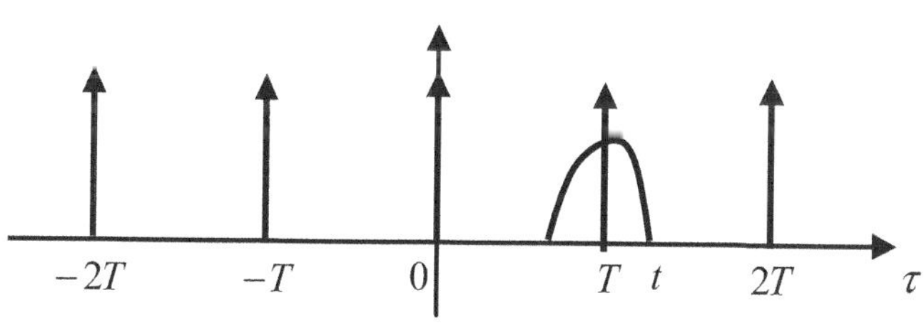

$$f_1(t) * f_2(t) = \int_{-\infty}^{+\infty} \delta(\tau - T) f_1(t-\tau) d\tau = f_1(t-T)$$

Therefore, in most cases: $f_1(t) * f_2(t) = \sum_{k=-\infty}^{k=+\infty} f_1(t-kT)$

This result can be obtained using Fourier transforms:

$$\begin{cases} f_1(t) \overset{TF}{\leftrightarrow} F_1(j\omega) \\ f_2(t) \overset{TF}{\leftrightarrow} F_2(j\omega) = \omega_0 \sum_{k=-\infty}^{k=+\infty} \delta(\omega - k\omega_0) \qquad \omega_0 = \dfrac{2\pi}{T} \end{cases}$$

$$f_1(t) * f_2(t) \overset{TF}{\leftrightarrow} F_1(j\omega) F_2(j\omega) = \omega_0 \sum_{k=-\infty}^{k=+\infty} F_1(j\omega)\delta(\omega - k\omega_0)$$

$$= \omega_0 \sum_{k=-\infty}^{k=+\infty} F_1(jk\omega_0)\delta(\omega - k\omega_0)$$

This can be written:

$$f_1(t) * f_2(t) \overset{TF}{\leftrightarrow} 2\pi \sum_{k=-\infty}^{k=+\infty} \frac{F_1(jk\omega_0)}{T} \delta(\omega - k\omega_0)$$

As this is the Fourier transform of a periodic function, we can establish the following:

$$f_1(t) * f_2(t) = \sum_{k=-\infty}^{k=+\infty} f_1(t - kT)$$

4.3.5. *Exercise 4.5*

a) Applying the derivation theorem:

$$\frac{d}{dt}\left(f_1(t) * \phi(t)\right) = f(t) * \frac{d\phi(t)}{dt}$$

The equation can then be written as follows:

$$f(t) = f(t) * \frac{d\phi(t)}{dt}$$

The only solution is: $\dfrac{d\phi(t)}{dt} = \delta(t)$

Thus: $\phi(t) = u(t) + K$

b) We set $f(t) = f_1(t) + A\delta(t)$ in the general case where a Dirac impulse can exist at the origin of the function:

$$F(j\omega) = F_1(j\omega) + A$$

We apply the integration theorem:

$$F_1(j\omega) + A = \frac{1}{j\omega}\left(F_1(j\omega) + A\right)\Phi(j\omega)$$

Thus: $\Phi(j\omega) = j\omega$

The reverse Fourier transform of $j\omega$ is the derivative of the function with a direct transform that is *1*.

Therefore: $\phi(t) = \dfrac{d\delta(t)}{dt}$

4.3.6. *Exercise 4.6*

$$s(t) = f(t)p(t) \Rightarrow S(j\omega) = \frac{1}{2\pi}\left(F(j\omega) * P(j\omega)\right)$$

a) The spectrum of $s(t)$ when $p(t) = \cos\omega_0 t$

$$P(j\omega) = \pi\left(\delta(\omega - n\omega_0) + \delta(\omega + n\omega_0)\right)$$

From this, we establish the following:

$$S(j\omega) = \frac{1}{2}\left(F(\omega - n\omega_0) + F(\omega + n\omega_0)\right)$$

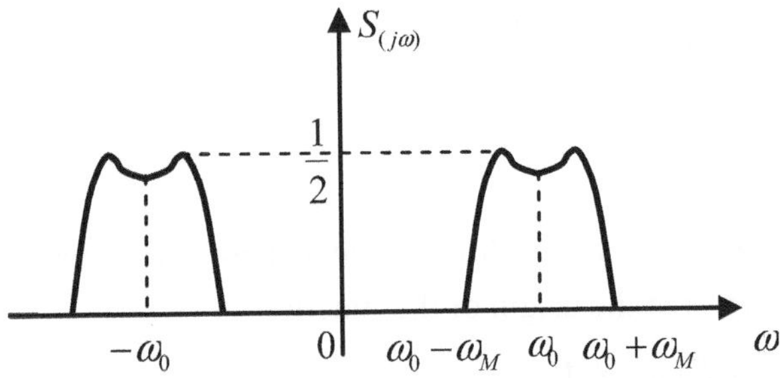

b) The spectrum of $s(t)$ when $p(t)=\sum_{n=-\infty}^{n=+\infty} Cr_{\frac{T}{2}}(t-nT)$

We can write $p(t)$ under the following form:

$$p(t)=Cr_{\frac{T}{2}}(t)\sum_{n=-\infty}^{n=+\infty}\delta(t-nT)$$

Thus: $P(j\omega)=\pi\sum_{n=-\infty}^{n=+\infty}\frac{\sin\frac{n\pi}{2}}{\frac{n\pi}{2}}\delta(\omega-n\omega_0)$

This allows us to set the following:

$$\boxed{S(j\omega)=\frac{1}{2}\sum_{n=-\infty}^{n=+\infty}\frac{\sin\frac{n\pi}{2}}{\frac{n\pi}{2}}F(\omega-n\omega_0)}$$

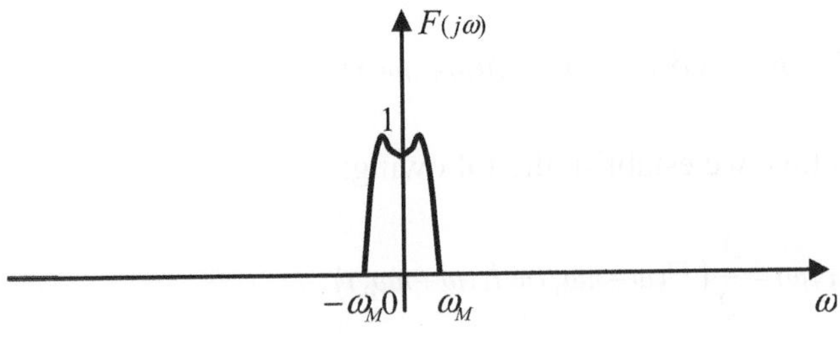

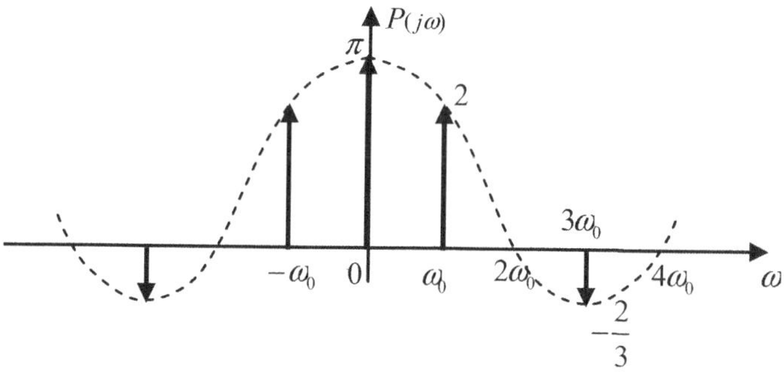

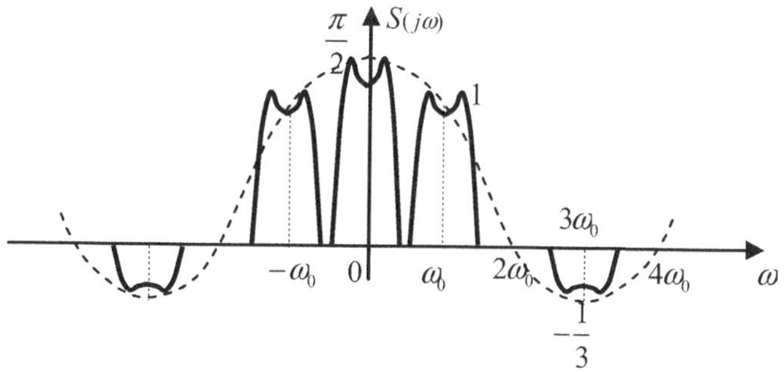

The output spectrum is therefore the spectrum of $f(t)$ repeated around the multiple frequencies of the fundamental frequency of $p(t)$ multiplied by a sine cardinal.

If condition $\omega > 2\omega_M$ is not verified, then there is spectral superposition.

c) The spectrum of $s(t)$ when $p(t)$ is any periodic function.

Function $p(t)$ can be decomposed into a Fourier series:

$$p(t) \overset{TF}{\leftrightarrow} P(j\omega) = 2\pi \sum_{n=-\infty}^{n=+\infty} C_n \delta(\omega - n\omega_0)$$

C_n is the coefficient of order n of the exponential Fourier decomposition.

We can then write:

$$S(j\omega) = \frac{1}{2\pi} F(j\omega) * 2\pi \sum_{n=-\infty}^{n=+\infty} C_n \delta(\omega - n\omega_0)$$
$$= \sum_{n=-\infty}^{n=+\infty} C_n F(\omega - n\omega_0)$$

The spectrum of $S(j\omega)$ corresponds to the spectrum of $F(j\omega)$ translated of $0, \pm\omega_0, \pm 2\omega_0$ and so on. The corresponding amplitudes having been multiplied by C_n.

In conclusion, when multiplying any function $f(t)$ with any other periodic function, the spectrum of the obtained signal corresponds to the spectrum of $f(t)$ shifted by $0, \pm\omega_0, \pm 2\omega_0, \ldots$ and the amplitudes of which are multiplied by the amplitudes of the spectral lines of the Fourier decomposition of the multiplication signal.

5

Correlation

5.1. Theoretical background

5.1.1. *Comparing signals*

Let us consider two real functions $f_1(t)$ and $f_2(t)$. If $f_1(t) = f_2(t)$ both functions are identical. When there is no relation between them whatsoever, they are said to be orthogonal. In other cases we can state that $f_1(t)$ contains a "certain proportion" of $f_2(t)$ represented by a coefficient C_{12}.

The similarities between these two function over an interval $[t_1, t_2]$ – in the case of a real function – can be written:

$$C_{12.} = \frac{\int_{t_1}^{t_2} f_1(t) f_2(t) dt}{\int_{t_1}^{t_2} f_2^2(t) dt} \quad \text{with} \quad \omega_0 = \frac{2\pi}{T} \qquad [5.2]$$

If $C_{12} = 1$ both functions are identical.

If $C_{12} = 0$ then both functions are orthogonal.

If we wish to compare the functions on the interval $]-\infty,+\infty[$, the comparison integral becomes:

$$\int_{-\infty}^{+\infty} f_1(t) f_2(t) dt$$

This type of comparison does present an inconvenience. Let us consider the two signals proposed in Figure 5.1.

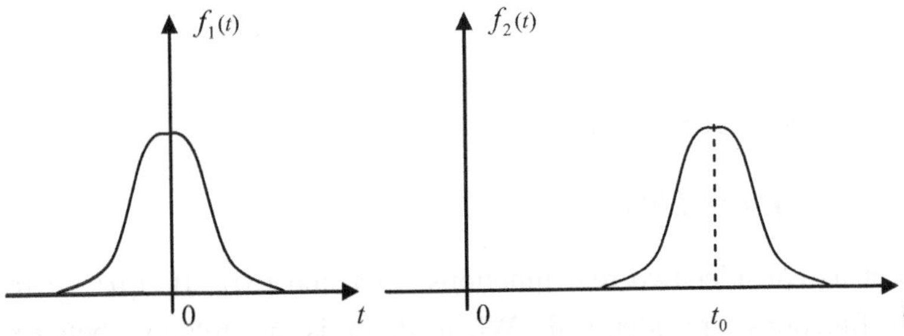

Figure 5.1. *Signal comparison*

These two signals are identical but for the fact that one has been subjected to a time-shift. However, the comparison integral is null.

In order to circumnavigate this difficulty, we will compare both functions shifted by a variable amount of time we will call τ. The comparison integral will then be a function of τ.

$$C(\tau) = \int_{-\infty}^{+\infty} f_1(t) f_2(t-\tau) dt$$

5.1.2. *Correlation function*

5.1.2.1. *Cross correlation function*

This function allows the comparison between two functions. It is written as follows:

$$C_{12}(\tau) = \int_{-\infty}^{+\infty} f_1(t) f_2(t-\tau)dt \qquad [5.1]$$

The shift performed on $f_2(t)$ can also be performed on $f_1(t)$ but in the opposite direction. We can then pose the correlation coefficient as:

$$C_{12}(t) = \int_{-\infty}^{+\infty} f_1(t+\tau) f_2(t)dt$$

We can also calculate the following:

$$C_{21}(\tau) = \int_{-\infty}^{+\infty} f_1(t-\tau) f_2(t)dt = \int_{-\infty}^{+\infty} f_1(t) f_2(t+\tau)dt$$

From this we establish the relation between the two coefficients:

$$C_{12}(\tau) = C_{21}(-\tau) \qquad [5.2]$$

– Case of complex functions:

$$C_{12}(\tau) = \int_{-\infty}^{+\infty} f_1(t) f_2^*(t-\tau)dt3 \qquad [5.3]$$

Hence: $C_{12}(\tau) = C_{21}^*(-\tau)$ [5.4]

5.1.2.2. *Autocorrelation function.*

This is the correlation of a function with itself.

$$C_1(\tau) = \int_{-\infty}^{+\infty} f_1(t) f_1^*(t-\tau)dt \qquad [5.5]$$

From this we establish: $C_{11}(\tau) = C_{11}^*(-\tau)$ [5.6]

In the case of real functions:

$$\boxed{\begin{array}{l} C_{11}(\tau) = \int_{-\infty}^{+\infty} f_1(t) f_1(t-\tau) dt \\ C_{11}(\tau) = C_{11}(-\tau) \end{array}} \qquad [5.7]$$

The autocorrelation function is thus an even function

5.1.2.3. *Graphical representation*

$C_{12}(\tau)$ represents the area between the time axis and the product of the functions after being subjected to the time-shift(s).

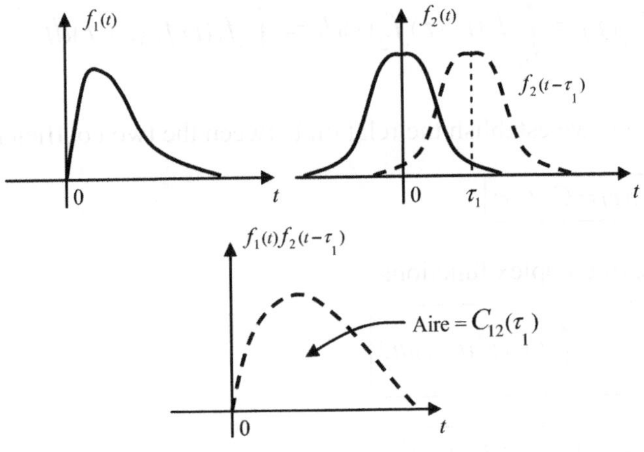

Figure 5.2. *Graphical interpretation*

5.1.3. *Properties of correlation functions*

5.1.3.1. *Correlation and convolution*

Consider two functions $f_1(t)$ and $f_2(t)$, the following is their convolution:

$$f_1(t) * f_2(t) = \int_{-\infty}^{+\infty} f_1(\tau) f_2(t-\tau) d\tau$$

The cross correlation $C_{12}(\tau)$ appears as follows:

$$C_{12}(\tau) = \int_{-\infty}^{+\infty} f_1(t) f_2(t-\tau) dt$$

Let us calculate the convolution of $f_1(t) * f_2(-t)$:

$$f_1(t) * f_2(-t) = \int_{-\infty}^{+\infty} f_1(\tau) f_2(\tau - t) d\tau$$

Hence the following result:

$$f_1(t) * f_2(-t)\Big|_{t=\tau} = C_{12}(\tau) \quad [5.8]$$

If $f_2(t)$ is even, the operations for the convolution and the correlation are the same.

5.1.3.2. *Extremums of correlation functions*

We obtain the maximal value of the autocorrelation function with $\tau = 0$:

$$\boxed{C_{11}(0) > |C_{11}(\tau)| \quad \forall \tau \neq 0} \quad [5.9]$$

We can also demonstrate the following results:

$$\boxed{\lim_{|\tau| \to \infty} \left(C_{11}(\tau) \right) = 0} \quad [5.10]$$

Schwartz's inequality $\boxed{C_{11}(0) C_{22}(0) \geq |C_{12}(\tau)|^2}$ [5.11]

5.1.3.3. *Correlation for signals with infinite energy*

Consider a signal with limitless energy:

$$\int_{-\infty}^{+\infty} f^2(t) dt \to \infty$$

The correlation functions previously defined no longer exist. We must use these average correlation functions defined by:

$$C_{11}(\tau) = \lim_{T\to\infty}\left(\frac{1}{T}\int_{-\frac{T}{2}}^{+\frac{T}{2}} f_1(t)f_1(t-\tau)dt\right)$$

$$C_{12}(\tau) = \lim_{T\to\infty}\left(\frac{1}{T}\int_{-\frac{T}{2}}^{+\frac{T}{2}} f_1(t)f_2(t-\tau)dt\right) \quad [5.12]$$

– Cases with periodic functions.

Consider two periodic functions $f_1(t)$ and $f_2(t)$ with periods of T. Functions $f_1(t)f_1(t-\tau)$ and $f_1(t)f_2(t-\tau)$ are also periodic with the same period. The integrals are also periodic and we simply need to use the mean value over a single period.

$$\boxed{\begin{aligned} C_{11}(\tau) &= \frac{1}{T}\int_{-\frac{T}{2}}^{+\frac{T}{2}} f_1(t)f_1(t-\tau)dt \\ C_{12}(\tau) &= \frac{1}{T}\int_{-\frac{T}{2}}^{+\frac{T}{2}} f_1(t)f_2(t-\tau)dt \end{aligned}} \quad [5.13]$$

The correlation functions are thus periodic with the same period T.

When the functions have different periods, function $f_1(t)f_2(t-\tau)$ can only be periodic if the ratio of the two periods is a rational number.

$\frac{T_2}{T_1} = \frac{p}{q}$ with p and q being integer numbers, so that $\frac{p}{q} \in \mathbb{Q}$

The period for the cross-correlation function is then $pT_1 = qT_2$. In that case, it can be written as follows:

$$\boxed{C_{12}(\tau) = \frac{1}{pT_1} \int_{-\frac{pT_1}{2}}^{+\frac{pT_1}{2}} f_1(t) f_2(t-\tau) dt} \qquad [5.14]$$

5.1.3.4. *Fourier transform of the correlation functions*

Let $F_1(j\omega)$ and $F_2(j\omega)$ be the Fourier transforms of $f_1(t)$ and $f_2(t)$.

Let $F_{12}(j\omega)$ be the Fourier transform of $C_{12}(\tau)$ defined by:

$$F_{12}(j\omega) = \int_{-\infty}^{+\infty} C_{12}(\tau) e^{-j\omega\tau} d\tau$$

If we apply Plancherel's theorem:

$$f_1(t) * f_2(t) \xleftrightarrow{TF} F_1(j\omega) F_2(j\omega).$$

We get:

$$\begin{cases} C_{12}(\tau) \xleftrightarrow{TF} F_{12}(j\omega) = F_1(j\omega) F_2(-j\omega) \\ C_{21}(\tau) \xleftrightarrow{TF} F_{21}(j\omega) = F_1(-j\omega) F_2(j\omega) \\ C_{11}(j\omega) \xleftrightarrow{TF} F_{11}(j\omega) = F_1(j\omega) F_1(-j\omega) \end{cases}$$

In the case of a real function:

$$F_1(-j\omega) = F_1^*(j\omega) \Rightarrow C_{11}(\tau) \xleftrightarrow{TF} |F_1(j\omega)|^2$$

The Fourier transform for the autocorrelation is always non-negative.

5.1.4. *Energy of a signal*

The energy contained in a signal is represented by the value at the origin of the autocorrelation function.

If $f_1(t)$ is a square integrable function:

$$\boxed{C_{11}(0) = \int_{-\infty}^{+\infty} f_1^2(t)dt} \qquad [5.15]$$

We apply the Parseval theorem to show the energy of a signal under the following form:

$$\frac{1}{2\pi}\int_{-\infty}^{+\infty} \left|F_1(j\omega)\right|^2 d\omega$$

We can define the spectral density of the energy of a signal:

$$\boxed{S_1(j\omega) = \frac{1}{2\pi}\left|F_1(j\omega)\right|^2} \qquad [5.16]$$

From this, we immediately establish that:

$$\boxed{C_{11}(\tau) \xleftrightarrow{TF} 2\pi S_1(j\omega)} \qquad [5.17]$$

This is the Wiener–Khinchin theorem:

The Fourier transform of the autocorrelation function of a signal is equal to 2π times the spectral density of the energy.

In a situation where the functions are not square integrable – the way that signals with infinite energy are – we apply the notion of power defined as follows:

$$\boxed{C_{11}(0) = \lim_{T\to\infty}\frac{1}{T}\int_{-\frac{T}{2}}^{+\frac{T}{2}} f_1^2(t)dt} \qquad [5.18]$$

Thus: $\boxed{C_{11}(\tau) \xleftrightarrow{TF} \lim_{T\to\infty} \frac{1}{T}\left|F_1(j\omega)\right|^2}$ [5.19]

$F_1(j\omega)$ is the Fourier transform of function $f_1(t)$ defined on the interval $\left[-\frac{T}{2}, +\frac{T}{2}\right]$.

In this case, the spectral power density is defined by:

$$S_1(j\omega) = \frac{1}{2\pi} F_1(j\omega)$$

5.2. Exercises

The exercises presented in this chapter come from B.P. Lathi [LAT 66].

5.2.1. *Exercise 5.1*

Determine the autocorrelation function of the following functions:

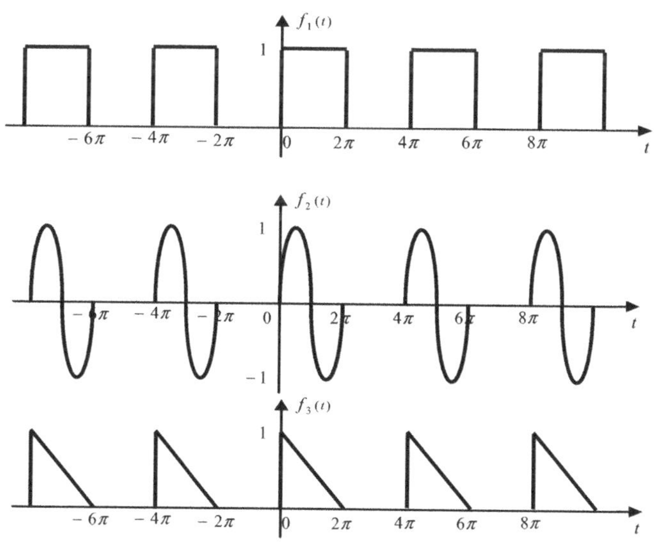

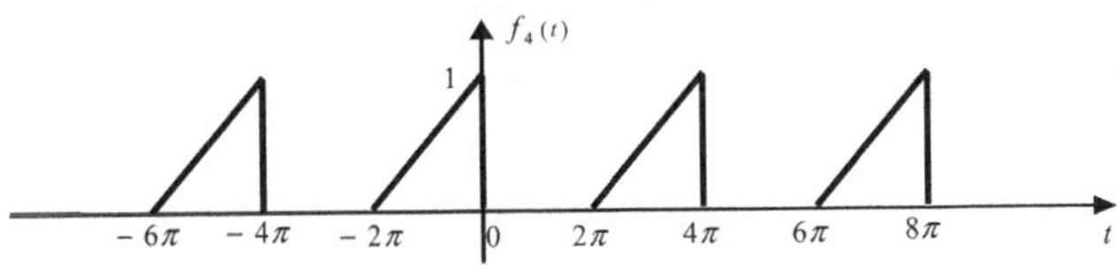

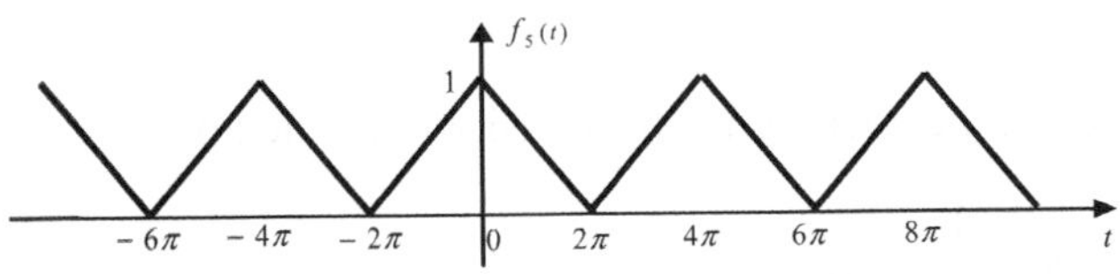

5.2.2. *Exercise 5.2*

Determine the autocorrelation functions $C_{13}(\tau)$, $C_{14}(\tau)$, $C_{15}(\tau)$ and $C_{34}(\tau)$ using the functions of exercise 5.1.

5.2.3. *Exercise 5.3*

If $\delta(t)$ is the Dirac impulse, demonstrate that the correlation of a function $f(t)$ by $\delta(t)$ gives us $f(\tau)$.

Demonstrate that the correlation of $f(t)$ by $\delta(t-t_0)$ gives us $f(\tau+t_0)$.

5.2.4. *Exercise 5.4*

We write $C_{12}^{(n)}(\tau)$ the nth derivative of $C_{12}(\tau)$.

Demonstrate that:

$$C_{12}^{(i+j)}(\tau)=(-1)^j\int_{-\infty}^{+\infty} f_1^{(i)}(t)f_2^{(j)}(t-\tau)dt$$

5.2.5. *Exercice 5.5*

Using the results of exercises 5.3 and 5.4, determine the cross-correlation of the following functions:

a)

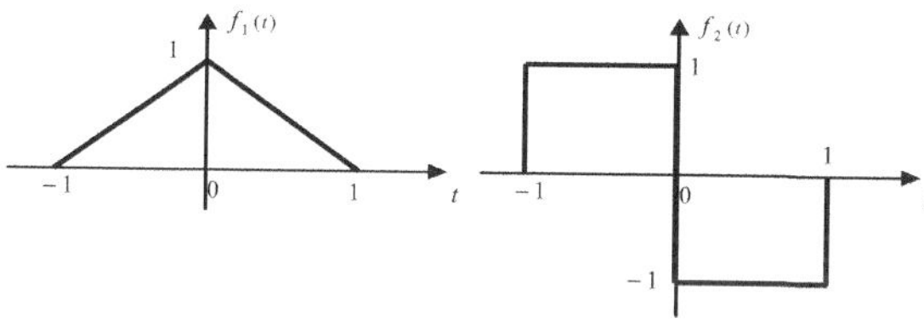

b)

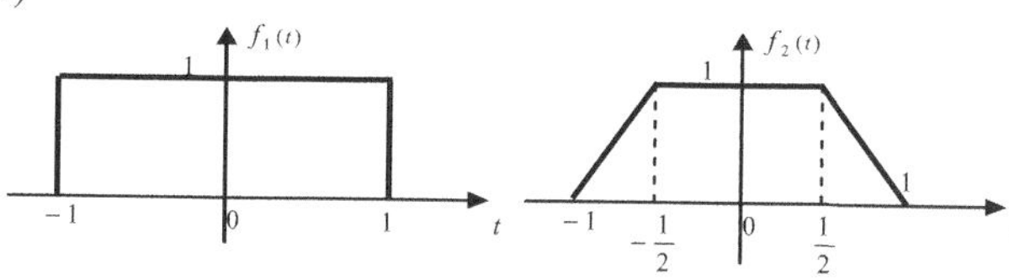

5.2.6. *Exercice 5.6*

Determine the cross-correlation of the following functions:

$$\begin{cases} f_1(t) = A\sin\omega_0 t + B\sin 2\omega_0 t \\ f_2(t) = C\sin\omega_0 t \end{cases}$$

5.2.7. *Exercise 5.7*

Let $x(t)$ be the following function:

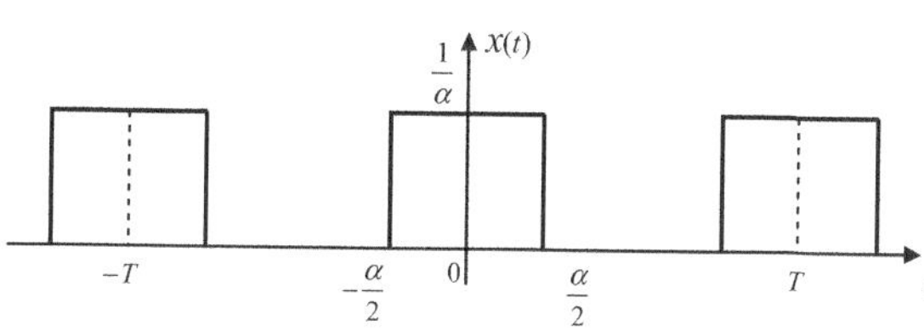

Determine:

a) its spectral amplitude X_n

b) its amplitude spectral density $X(j\omega)$

c) its power spectrum

d) autocorrelation $C_{xx}(\tau)$. Develop this function into a Fourier series

e) calculate the mean power P of signal $x(t)$

5.2.8. *Exercice 5.8.*

For the time-scale functions described below, determine, $S(j\omega)$ and $C_{ff}(\tau)$:

a) $f(t) = \delta(t)$

b) $f(t) = \dfrac{1}{\Delta} Cr_{\Delta}(t)$

c) $f(t) = e^{-\alpha t} u(t)$

d) $f(t) = 2\delta(t) + \delta(t-1)$

5.2.9. *Exercise 5.9*

The spectral density of energy $S(j\omega)$ of time function $f(t)$ is the following:

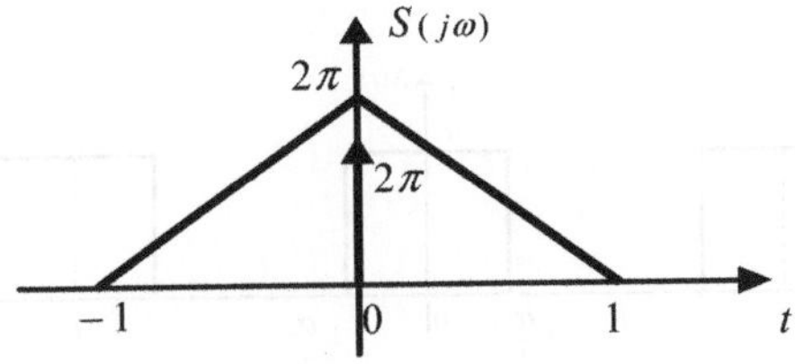

a) Calculate and plot the autocorrelation function of the signal

b) Calculate the mean direct (DC) power of the signal

c) Calculate the "alternating (AC)" power of $f(t)$

5.2.10. *Exercise 5.10*

Consider the following filter:

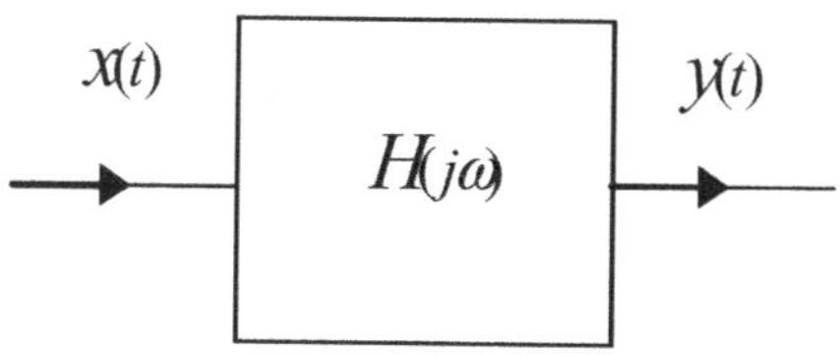

This type of filter is said to be "matched" if $H(j\omega) = X^*(j\omega)$.

a) What is the relationship tying the input value and the impulse response of the filter?

b) Demonstrate that the output value $y_{(t)}$ represents the autocorrelation function of the input value for which the filter is adapted.

5.2.11. *Exercise 5.11*

We wish to establish a correlator to calculate the cross-correlation between any function $f(t)$ and function $e^{at}u(-t)$.

$f(t)$ is the system input. If we wish for the output to represent the cross-correlation function:

a) What must be the impulse response from the circuit?

b) What will its transfer function then be?

c) With $a = 1$, how can we realize this circuit?

5.2.12. Exercise 5.12

Demonstrate that if $f_1(t)$ and $f_2(t)$ are two complex functions of t, then:

$$C_{12}(\tau) \xleftrightarrow{TF} F_1(j\omega)F_2^*(j\omega)$$
$$C_{21}(\tau) \xleftrightarrow{TF} F_1^*(j\omega)F_2(j\omega)$$

Demonstrate that:

$$\int_{-\infty}^{+\infty} f_1(t)f_2^*(t)dt = \frac{1}{2\pi}\int_{-\infty}^{+\infty} F_1(j\omega)F_2^*(j\omega)d\omega$$
$$\int_{-\infty}^{+\infty} f_1(t)f_2^*(-t)dt = \frac{1}{2\pi}\int_{-\infty}^{+\infty} F_1(j\omega)F_2^*(-j\omega)d\omega$$

What happens to these relations when both functions are real?

5.2.13. Exercise 5.13

If $f_2(t)$ is the response of a linear system to signal $f_1(t)$, demonstrate that:

$$C_{21}(\tau) = C_{11}(\tau) * h(\tau)$$
$$C_{22}(\tau) = \int_{-\infty}^{+\infty} C_{hh}(t)C_{11}(\tau - t)dt$$

$$\text{with} \begin{cases} h(t) = \text{ impulse response from the system} \\ C_{hh}(\tau) = \text{ auto-correlation function of } h(t) \end{cases}$$

Demonstrate that:

$$TF\left(C_{22}(\tau)\right) = \left|H(j\omega)\right|^2 TF\left(C_{11}(\tau)\right)$$

5.2.14. *Exercise 5.14*

Consider $f(t) = \sin at + \sin bt$.

a) Does $f(t)$ represent a finite energy signal?

b) Calculate $C_{ff}(\tau)$ and $\lim_{\tau\to\infty} C_{ff}(\tau)$

c) Calculate $C_{ff}(\tau)$ where $a = 2$ and $b = 5$ then with $a = 1$ and $b = \sqrt{2}$

5.3. Solutions to the exercises

5.3.1. *Exercise 5.1*

a) Calculation for $C_{11}(\tau)$

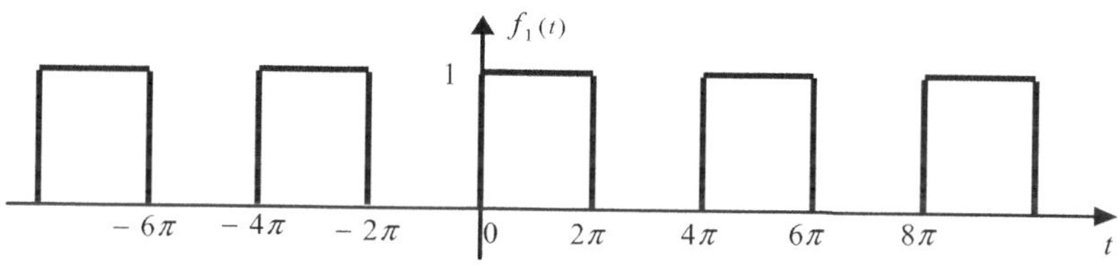

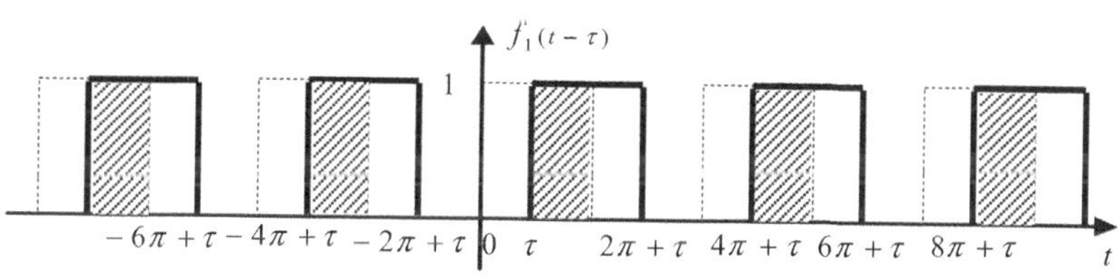

Consider the interval over a period $[-2\pi, +2\pi]$ and a shift of $\tau \in [0, 2\pi]$.

We write $f_1(t)$: $\begin{cases} f_1(t) = 0 & t \in [-2\pi, 0] \\ f_{(t)1} = 1 & t \in [0, +2\pi] \end{cases}$

We write $f_1(t-\tau)$: $\begin{cases} f_1(t-\tau)=1 & t\in[-2\pi,-2\pi+\tau] \\ f_1(t-\tau)=0 & t\in[-2\pi+\tau,\tau] \\ f_1(t-\tau)=1 & t\in[\tau,+2\pi] \end{cases}$

The following is the autocorrelation function:

$$C_{11}(\tau)=\frac{1}{4\pi}\int_{-2\pi}^{+2\pi} f_1(t)f_1(t-\tau)dt$$

$$=\frac{1}{4\pi}\int_{\tau}^{+2\pi} dt$$

$$=\frac{1}{2}-\frac{\tau}{4\pi}$$

As this autocorrelation function is even, the part corresponding to $\tau<0$ can be obtained through symmetry.

The following is the expression of the autocorrelation function over a single period:

$$\boxed{C_{11}(\tau)=\frac{1}{2}-\frac{|\tau|}{4\pi}}$$

The autocorrelation function is periodic with a period of 4π

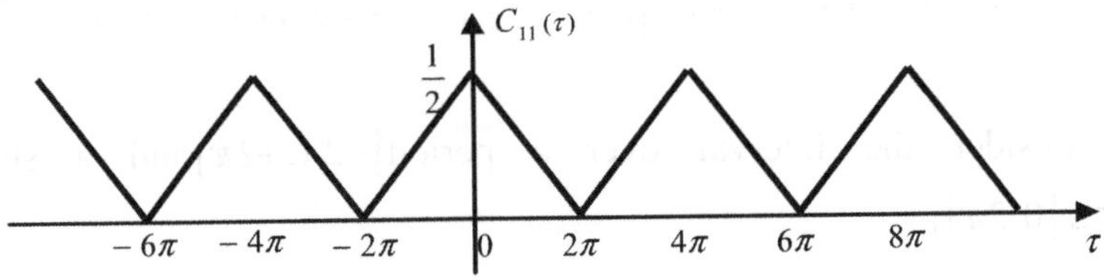

b) Calculation of $C_{22}(\tau)$ for $\tau>0$

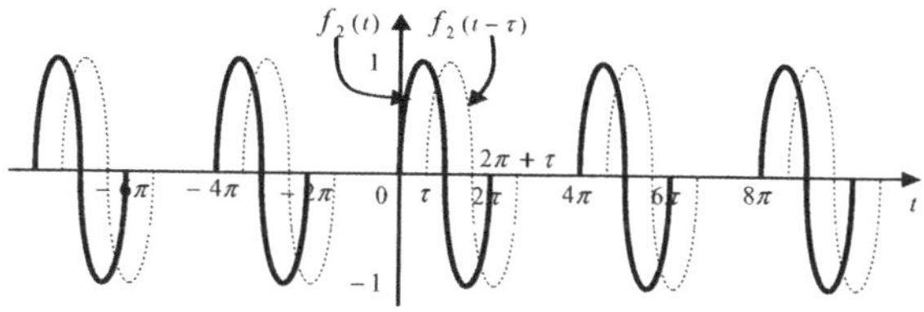

Consider the interval over a period $[-2\pi,+2\pi]$ and a shift of $\tau \in [0,2\pi]$.

We write $f_2(t)$: $\begin{cases} f_2(t)=0 & t\in[-2\pi,0] \\ f_2(t)=\sin t & t\in[0,+2\pi] \end{cases}$

We write $f_2(t-\tau)$: $\begin{cases} f_2(t-\tau)=\sin(t-\tau) & t\in[-2\pi,-2\pi+\tau] \\ f_2(t-\tau)=0 & t\in[-2\pi+\tau,\tau] \\ f_2(t-\tau)=\sin(t-\tau) & t\in[\tau,+2\pi] \end{cases}$

The following is the autocorrelation function:

$$C_{22}(\tau)=\frac{1}{4\pi}\int_{-2\pi}^{+2\pi} f_2(t)f_2(t-\tau)dt$$
$$=\frac{1}{4\pi}\int_{\tau}^{+2\pi}\sin t\sin(t-\tau)dt$$

After integration:

$$\boxed{C_{22}(\tau)=\frac{1}{8\pi}\left(\sin\tau+(2\pi-\tau)\cos\tau\right)}$$

$C_{22}(-\tau)$ can be obtained through symmetry

$$C_{22}(\tau)=\frac{1}{8\pi}\left(\sin|\tau|+(2\pi-|\tau|)\cos\tau\right)$$

The autocorrelation function is periodic with a period of 4π

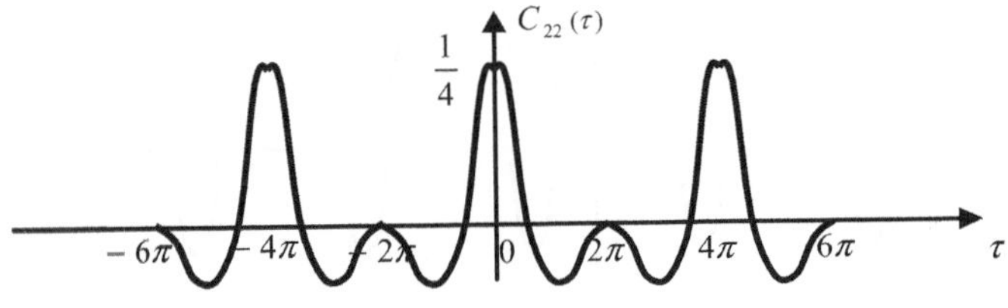

c) Calculation of $C_{33}(\tau)$ for $\tau > 0$

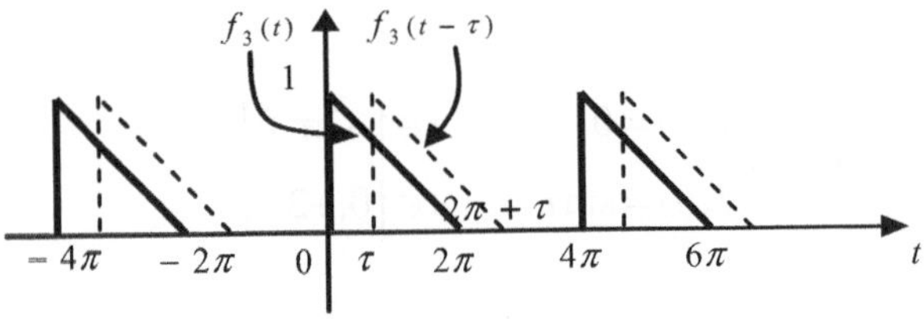

The autocorrelation function is periodic with a period of 4π.

Consider the interval over a period $[-2\pi,+2\pi]$ and a shift of $\tau \in [0,2\pi]$.

We write $f_3(t)$: $\begin{cases} f_3(t) = 0 & t \in [-2\pi, 0] \\ f_3(t) = 1 - \dfrac{t}{2\pi} & t \in [0,+2\pi] \end{cases}$

We write $f_3(t-\tau)$: $\begin{cases} f_3(t-\tau) = -1 - \dfrac{t-\tau}{2\pi} & t \in [-2\pi, -2\pi+\tau] \\ f_3(t-\tau) = 0 & t \in [-2\pi+\tau, \tau] \\ f_3(t-\tau) = 1 - \dfrac{t-\tau}{2\pi} & t \in [\tau, +2\pi] \end{cases}$

The autocorrelation function becomes:

$$C_{33}(\tau) = \frac{1}{4\pi}\int_{-2\pi}^{+2\pi} f_3(t) f_3(t-\tau)dt$$

$$= \frac{1}{4\pi}\int_{\tau}^{+2\pi}\left(1 - \frac{t}{2\pi}\right)\left(1 - \frac{t-\tau}{2\pi}\right)dt$$

After integration:

$$C_{33}(\tau)=\frac{1}{6}\left(1+\frac{\tau}{4\pi}\right)\left(1-\frac{\tau}{2\pi}\right)^2$$

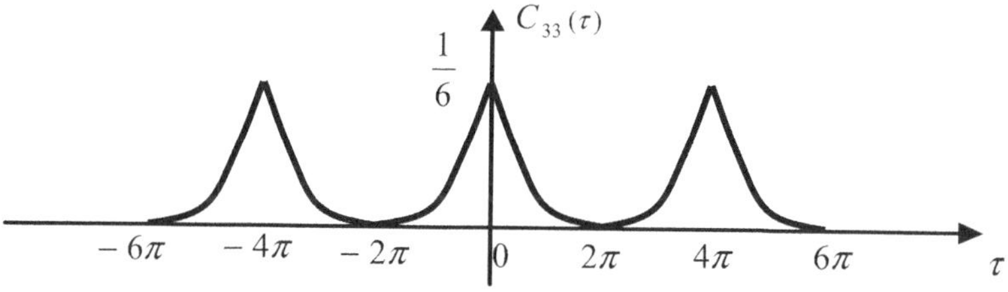

$C_{33}(-\tau)$ can again be obtained through symmetry.

Hence the following result:

$$C_{33}(\tau)=\frac{1}{6}\left(1+\frac{|\tau|}{4\pi}\right)\left(1-\frac{|\tau|}{2\pi}\right)^2$$

d) Calculation of $C_{44}(\tau)$

We note that $f_4(t)=f_3(-t)$

Knowing that $C_{44}(\tau)=f_4(\tau)*f_4(-\tau)$, we get:

$$C_{44}(\tau)=f_4(\tau)*f_4(-\tau)=f_3(-\tau)*f_3(\tau)=C_{33}(\tau)$$

$$C_{44}(\tau)=C_{33}(\tau)=\frac{1}{6}\left(1+\frac{\tau}{4\pi}\right)\left(1-\frac{\tau}{2\pi}\right)^2 \quad \text{where } \tau>0$$

The autocorrelation function is periodic with a period of 4π.

$$C_{44}(\tau)=C_{33}(\tau)=\frac{1}{6}\left(1+\frac{|\tau|}{4\pi}\right)\left(1-\frac{|\tau|}{2\pi}\right)^2$$

e) Calculation of $C_{55}(\tau)$

By performing a calculation analogous to the previous one, we obtain, for a shift of $\tau \in [0, 2\pi]$.

$$\boxed{C_{55}(\tau) = \frac{|\tau|^3}{24\pi^3} - \frac{\tau^2}{8\pi^2} + \frac{1}{3}}$$

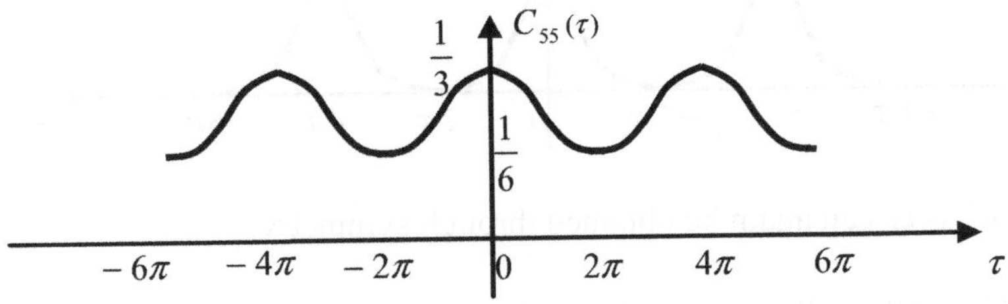

The autocorrelation function is periodic with a period of 4π

5.3.2. *Exercice 5.2*

– Calculation of $C_{13}(\tau)$

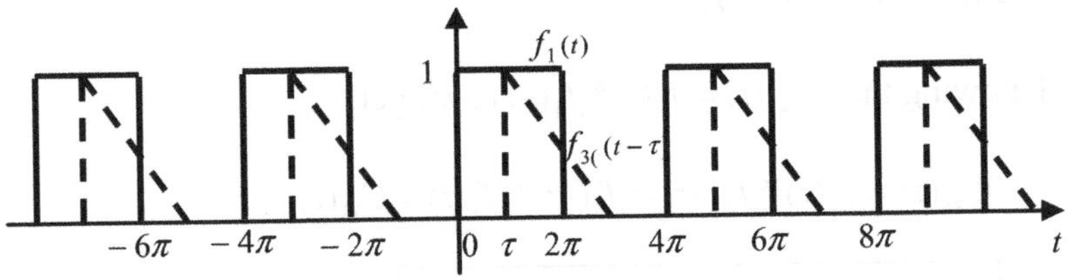

Consider the interval over a period $[-2\pi, +2\pi]$ and a shift of $\tau \in [0, 2\pi]$.

The functions have been defined in the various intervals during the first exercise in this chapter.

We can then calculate the cross-correlation function:

$$C_{13}(\tau)=\frac{1}{4\pi}\int_{2\pi}^{+2\pi} f_1(t)f_3(t-\tau)dt$$

$$=\frac{1}{4\pi}\int_{\tau}^{+2\pi}\left(1-\frac{t-\tau}{2\pi}\right)dt$$

Thus, after calculating the integral:

$$\boxed{C_{13}(\tau)=\frac{1}{4}-\left(\frac{\tau}{4\pi}\right)^2 \qquad \tau\in[0,2\pi]}$$

When $\tau\in[-2\pi,0]$, the integration limits are changed. We get:

$$C_{13}(\tau)=\frac{1}{4\pi}\int_{0}^{2\pi+\tau}\left(1-\frac{t-\tau}{2\pi}\right)dt$$

Thus, after calculating the integration:

$$\boxed{C_{13}(\tau)=\frac{1}{4}+\frac{\tau}{4\pi}+\left(\frac{\tau}{4\pi}\right)^2 \qquad \tau\in[-2\pi,0]}$$

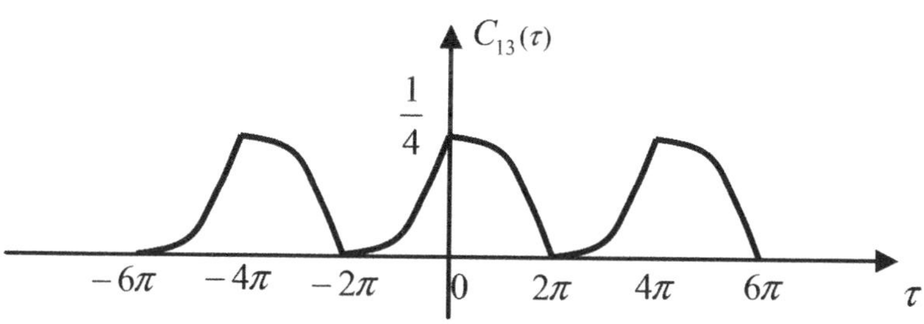

– Calculation of $C_{14}(\tau)$

The method of calculation is the same as for $C_{13}(\tau)$. We get:

$$C_{14}(\tau) = \frac{\tau}{4\pi} - \left(\frac{\tau}{4\pi}\right)^2 \qquad \tau \in [0, 2\pi]$$

$$C_{13}(\tau) = \left(\frac{\tau}{4\pi}\right)^2 \qquad \tau \in [-2\pi, 0]$$

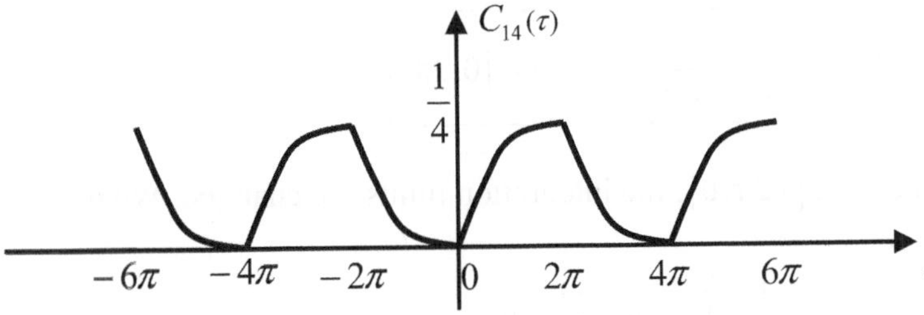

– Calculation of $C_{15}(\tau)$

The method of calculation is the same as in previous cases. We get:

$$C_{15}(\tau) = \frac{1}{4} + \frac{\tau}{4\pi} - \frac{\tau^2}{8\pi^2} \qquad \tau \in [0, 2\pi]$$

$$C_{15}(\tau) = \frac{1}{4} + \frac{\tau}{4\pi} + \frac{\tau^2}{8\pi^2} \qquad \tau \in [-2\pi, 0]$$

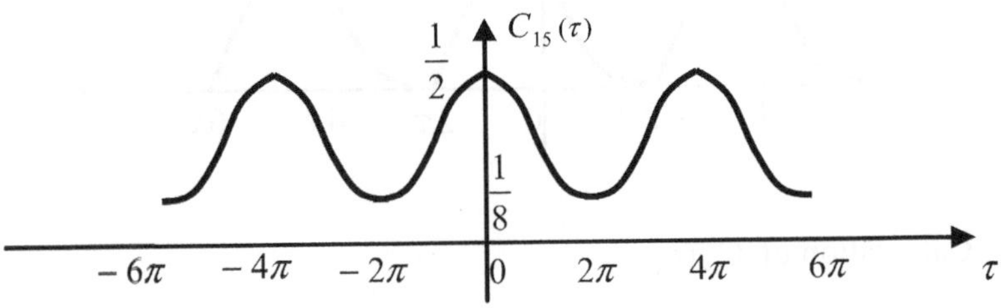

– Calculation for $C_{34}(\tau)$

The method of calculation is the same as in previous cases. We get:

$$C_{34}(\tau) = \frac{1}{4\pi}\left(\tau - \frac{\tau^2}{2\pi} + \frac{\tau^3}{24\pi^2}\right) \qquad \tau \in [0, 2\pi]$$

$$C_{34}(\tau) = -\frac{1}{4\pi}\frac{\tau^3}{24\pi^2} \qquad \tau \in [-2\pi, 0]$$

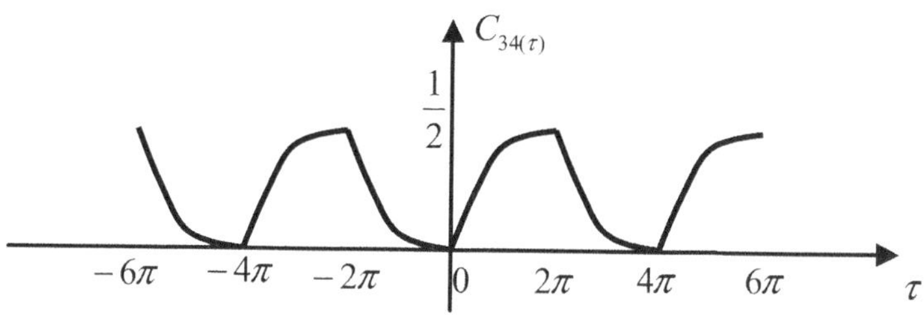

REMARK.–

The calculation of $C_{55}(\tau)$ can be performed by noting that $f_5(\tau) = f_3(\tau) + f_4(\tau)$

From this we establish that:

$$\begin{aligned} C_{55}(\tau) &= C_{33}(\tau) + C_{44}(\tau) + C_{34}(\tau) + C_{43}(\tau) \\ &= C_{33}(\tau) + C_{44}(\tau) + C_{34}(\tau) + C^{*}_{34}(-\tau) \end{aligned}$$

5.3.3. *Exercise 5.3*

The correlation of $f(t)$ by a Dirac impulse appears as follows:

$$C(\tau) = \int_{-\infty}^{+\infty} f(t)\delta(t-\tau)dt$$

The function we are looking to ingrate is null except when $t=\tau$, therefore:

$$C(\tau)=f(\tau)\int_{-\infty}^{+\infty}\delta(t-\tau)dt=f(\tau)$$

As such:

$$C(\tau)=\int_{-\infty}^{+\infty}f(t)\delta(t-t_0-\tau)dt=\int_{-\infty}^{+\infty}f(t)\delta(t-(t_0+\tau))dt$$

$$=f(t_0+\tau)\int_{-\infty}^{+\infty}\delta(t)dt=f(t_0+\tau)$$

5.3.4. *Exercice 5.4*

If $P(\tau)$ is the convolution of two functions $f_1(\tau)$ and $g(\tau)$, the properties of derivation of this convolution allow us to state the following:

$$P^{(i+j)}(\tau)=f_1^{(i)}(\tau)*g^{(j)}(\tau)$$

We set $g(\tau)=f_2(-\tau)$ then: $g^{(j)}(\tau)=(-1)^j f_2^{(j)}(-\tau)$

In this case: $P(\tau)=C_{12}(\tau)$

From this we establish that:

$$P^{(i+j)}(\tau)=C_{12}^{(i+j)}(\tau)=(-1)^j f_1^{(i)}(\tau)*g^{(j)}(-\tau)$$

This can be posed as an integral:

$$\boxed{C_{12}^{(i+j)}(\tau)=(-1)^j\int_{-\infty}^{+\infty}f_1^{(i)}(\tau)f_2^{(j)}(t-\tau)dt}$$

5.3.5. *Exercise 5.5*

a) Let us determine the derivative of $f_2(t)$:

$$f_2^{'}(t) = \delta(t) - 2\delta(t) + \delta(t-1)$$

By applying the results obtained in exercise 5.3, we can write the cross-correlation function between $f_1(t)$ and $f_2^{'}(t)$ under the following form:

$$C_{12^{'}}(\tau) = f_1(\tau+1) - 2f_1(\tau) + f_1(\tau-1)$$

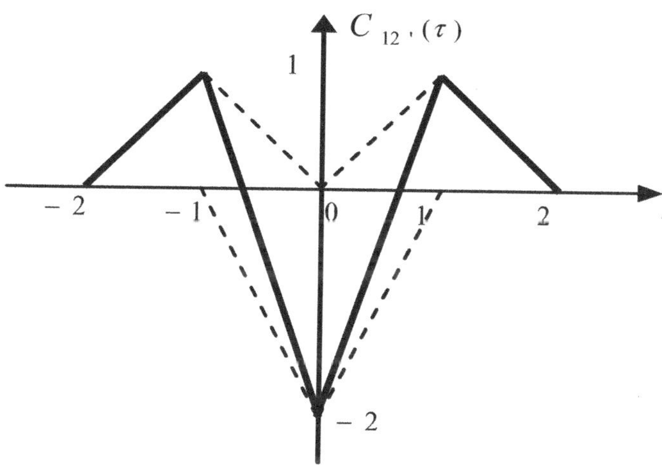

By applying the results from exercise 5.4, we can state the following:

$$C_{12}^{'}(\tau) = -f_1(\tau) * f_2^{'}(\tau) = -C_{12^{'}}(\tau)$$

Thus: $C_{12}(\tau) = -\int_{-\infty}^{\tau} C_{12^{'}}(x)dx$

Hence: $C_{12}(\tau) = -\int_{-\infty}^{\tau} \left(f_1(x+1) - 2f_1(x) + f_1(x-1)\right)dx$

– $\tau < -2$ and $\tau > 2$

$$\boxed{C_{12}(\tau) = 0}$$

– $-2 < \tau < -1$

$$C_{12(\tau)} = -\int_{-2}^{\tau} (x+2)dx = -\frac{\tau^2}{2} - 2\tau - 2$$

$$\boxed{C_{12}(\tau) = -\frac{\tau^2}{2} - 2\tau - 2}$$

– $-1 < \tau < 0$

$$C_{12}(\tau) = -\int_{-2}^{-1} (x+2)dx - \int_{-1}^{\tau} (-3x-2)dx = \frac{3\tau^2}{2} + 2\tau$$

$$\boxed{C_{12}(\tau) = \frac{3\tau^2}{2} + 2\tau}$$

– $0 < \tau < 1$

$$C_{12}(\tau) = -\int_{-2}^{-1} (x+2)dx - \int_{-1}^{2} (-3x-2)dx - \int_{0}^{\tau} (3x-2)dx = -\frac{3\tau^2}{2} + 2\tau$$

$$\boxed{C_{12}(\tau) = -\frac{3\tau^2}{2} + 2\tau}$$

– $1 < \tau < 2$

$$C_{12}(\tau)=-\int_{-2}^{-1}(x+2)dx-\int_{-1}^{2}(-3x-2)dx-\int_{0}^{1}(3x-2)dx-\int_{1}^{\tau}(-x+2)dx=\frac{\tau^2}{2}-2\tau+2$$

$$\boxed{C_{12}(\tau)=\frac{\tau^2}{2}-2\tau+2}$$

This explains the following plot:

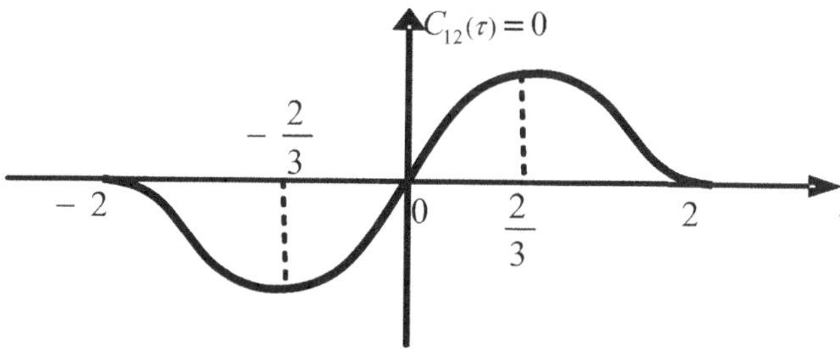

b) The method of calculation is the same as in case *a*. We obtain the following results:

– $\tau<-2$ and $\tau>2$

$$\boxed{C_{12}(\tau)=0}$$

c) $-2<\tau<-\frac{3}{2}$

$$\boxed{C_{12}(\tau)=(\tau+2)^2}$$

d) $-\frac{3}{2}<\tau<-\frac{1}{2}$

$$\boxed{C_{12}(\tau)=\tau+\frac{7}{4}}$$

e) $-\frac{1}{2} < \tau < \frac{1}{2}$

$$\boxed{C_{12}(\tau) = \frac{3}{2} - \tau^2}$$

f) $\frac{1}{2} < \tau < \frac{3}{2}$

$$\boxed{C_{12}(\tau) = -\tau + \frac{7}{4}}$$

g) $\frac{3}{2} < \tau < 2$

$$\boxed{C_{12}(\tau) = (\tau - 2)^2}$$

Thus the following plot:

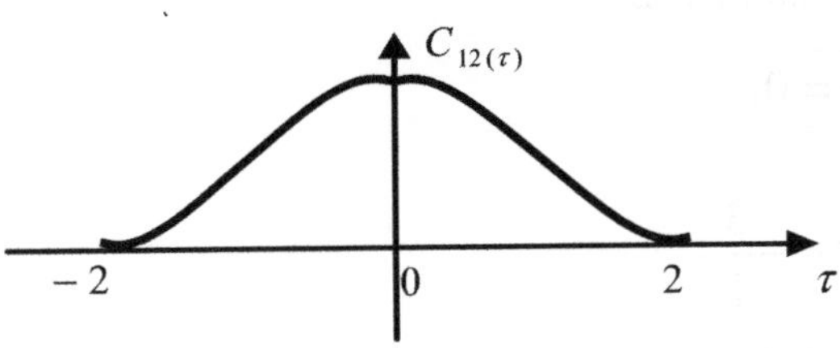

5.3.6. *Exercise 5.6*

The period of $f_1(t)$ is equal to that of $f_2(t)$. Integration will take place over a period of $T = \frac{2\omega}{\omega_0}$.

$$C_{12}(\tau) = \frac{1}{T} \int_{-\frac{T}{2}}^{+\frac{T}{2}} f_1(t) f_2(t - \tau) dt$$

$$C_{12}(\tau)=\frac{1}{T}\int_{-\frac{T}{2}}^{+\frac{T}{2}} AC\sin\omega_0 t\sin\omega_0(t-\tau)dt+\frac{1}{T}\int_{-\frac{T}{2}}^{+\frac{T}{2}} BC\sin 2\omega_0 t\sin\omega_0(t-\tau)dt$$

Thus after calculation:

$$\boxed{C_{12}(\tau)=\frac{AC}{2}\cos\omega_0\tau}$$

5.3.7. *Exercise 5.7*

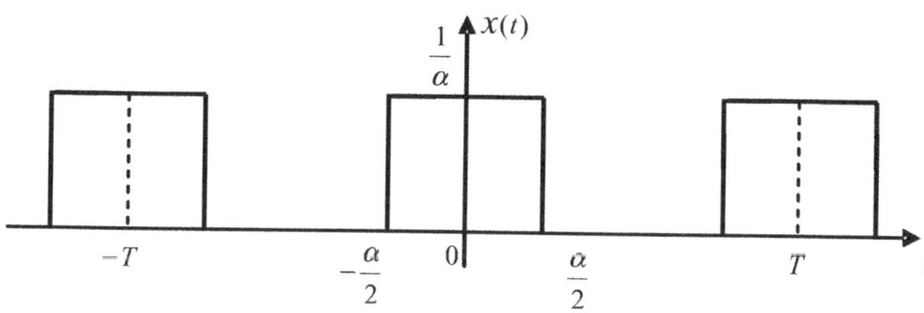

a) $x(t)$ is periodic and thus can be decomposed into a Fourier series.

$$X_n=\frac{1}{T}\int_{-\frac{T}{2}}^{+\frac{T}{2}} x(t)e^{-jn\omega_0 t}dt=\frac{1}{T}\int_{-\frac{\alpha}{2}}^{+\frac{\alpha}{2}}\frac{1}{\alpha}e^{-jn\omega_0 t}dt$$

Thus, after calculation: $\boxed{X_n=\frac{1}{T}\frac{\sin n\pi\frac{\alpha}{T}}{n\pi\frac{\alpha}{T}}}$

b) $x(t)$ also has a Fourier transform $X(j\omega)$.

$$x(t)\xleftrightarrow{TF}X(j\omega)$$

$$X(j\omega) = 2\pi \sum_{n=-\infty}^{n=+\infty} X_n \delta(\omega - n\omega_0)$$

Thus:

$$X(j\omega) = \omega_0 \sum_{n=-\infty}^{n=+\infty} \frac{\sin n\omega_0 \frac{\alpha}{2}}{n\omega_0 \frac{\alpha}{2}} \delta(\omega - n\omega_0)$$

c) The power spectrum is $|X_n|^2$.

$$S_n = |X_n|^2 = \frac{1}{T^2}\left(\frac{\sin n\pi \frac{\alpha}{T}}{n\pi \frac{\alpha}{T}}\right)^2$$

d) We compare $x(t)$ and $x(t-\tau)$ on interval $\left[-\frac{T}{2}, +\frac{T}{2}\right]$.

$$x(t) = \begin{cases} 0 & t \in \left[-\frac{T}{2}, -\frac{\alpha}{2}\right] \\ \frac{1}{\alpha} & t \in \left[-\frac{\alpha}{2}, +\frac{\alpha}{2}\right] \\ 0 & t \in \left[+\frac{\alpha}{2}, +\frac{T}{2}\right] \end{cases}$$

$$= \begin{cases} \frac{1}{\alpha} & t \in \left[-\frac{T}{2}, -T + \frac{\alpha}{2} + \tau\right] \\ 0 & t \in \left[-T + \frac{\alpha}{2} + \tau, -\frac{\alpha}{2} + \tau\right] \\ \frac{1}{\alpha} & t \in \left[-\frac{\alpha}{2} + \tau, +\frac{\alpha}{2} + \tau\right] \\ 0 & t \in \left[+\frac{\alpha}{2} + \tau, +\frac{T}{2}\right] \end{cases}$$

We must consider two cases according to value α :

$$-\frac{\alpha}{2} < \frac{T}{4}$$

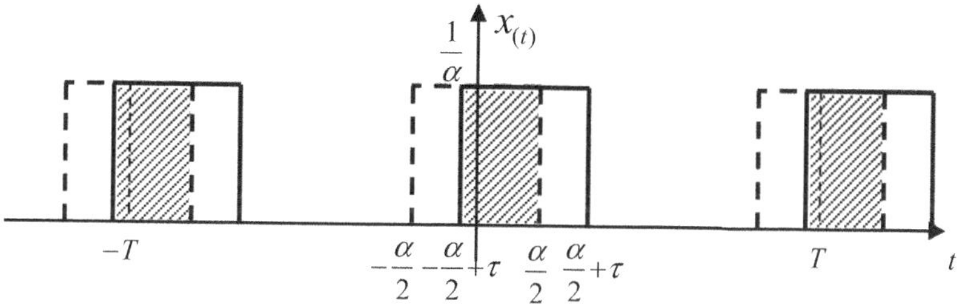

Both functions overlap only on interval $\left[-\frac{\alpha}{2}+\tau,+\frac{\alpha}{2}\right]$

thus $C_{11}(\tau) = \frac{1}{T}\int_{-\frac{\alpha}{2}+\tau}^{\frac{\alpha}{2}} \frac{1}{\alpha^2}dt = \frac{1}{\alpha T}\left(1-\frac{\tau}{\alpha}\right)$ for $\tau \in [0,\alpha]$

and $C_{11}(\tau) = 0$ for $\tau \in \left[\alpha, \frac{T}{2}\right]$

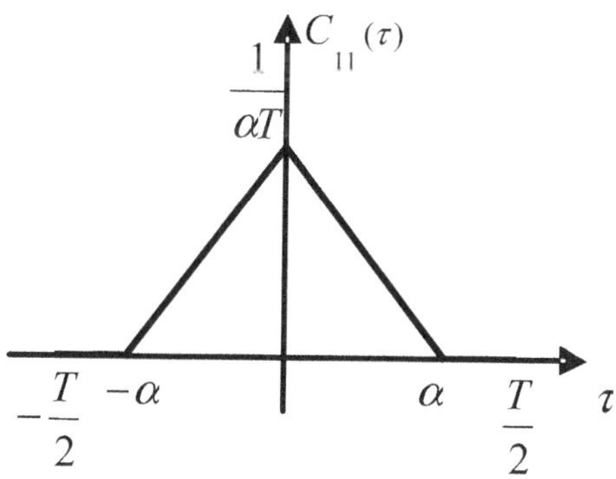

$$-\frac{T}{4}<\frac{\alpha}{2}<\frac{T}{2}$$

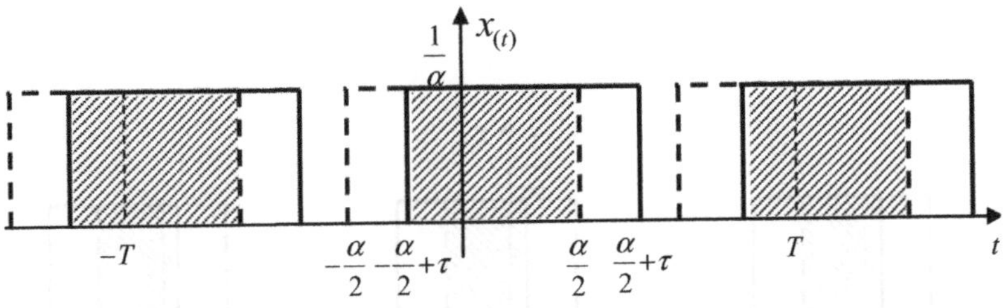

When $\frac{\alpha}{2}+\tau>T-\frac{\alpha}{2}$ there is complete overlap $\forall t$. Therefore if $\tau>T-\alpha$.

So if $0<\tau<T-\alpha$ $\quad C_{11}(\tau)=\frac{1}{\alpha T}\left(1-\frac{\tau}{\alpha}\right)$ as we have seen previously:

If $T-\alpha<\tau<\frac{T}{2}$ $\quad C_{11}(\tau)=\frac{1}{\alpha T}\left(2-\frac{T}{\alpha}\right)$

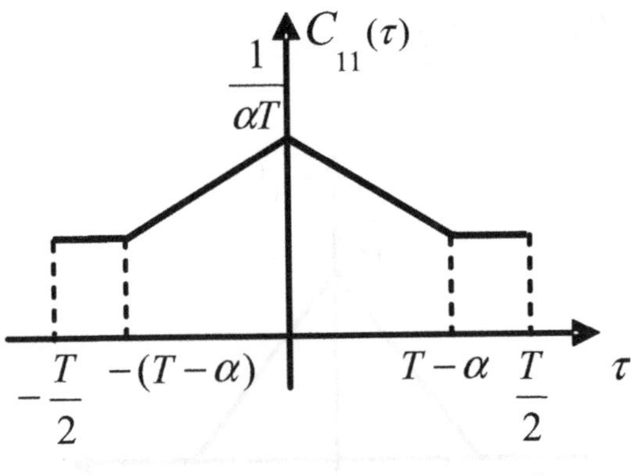

The Fourier series development is obtained by considering the parity of these functions:

$$F_n = \frac{4}{T}\int_0^{\frac{T}{2}} C_{11}(t)\cos n\omega_0 t dt$$

After this calculation, we obtain the same result in both cases:

$$\boxed{F_n = \frac{1}{T^2}\left(\frac{\sin n\omega_0 \frac{\alpha}{2}}{n\omega_0 \frac{\alpha}{2}}\right)^2}$$

e) The mean power of signal $x(t)$ is:

$$P = \frac{1}{T}\int_{-\frac{T}{2}}^{\frac{T}{2}} x^2(t)dt = \sum_{n=-\infty}^{n=+\infty} |X_n|^2$$

According to Parseval's theorem, we can write: $P = C_{11}(0)$

Thus: $\boxed{P = \frac{1}{\alpha T}}$

5.3.8. *Exercise 5.8*

a) $f(t) = \delta(t)$

The Fourier transform is $\delta(t) \xleftrightarrow{TF} 1$

Hence the spectral density modulus: $|F(j\omega)| = 1$

The spectral density of the energy is: $S(\omega)=\frac{1}{2\pi}\left|F(j\omega)\right|^2$

Hence: $S(\omega)=\frac{1}{2\pi}$

The autocorrelation function $C_{ff}(\tau)$ appears as follows:

$$C_{ff}(\tau)=\int_{-\infty}^{+\infty}\delta(t)\delta(t-\tau)dt=\delta(\tau)$$

The authors warn against this notation as the direct product of two Dirac distributions has no actual meaning on R.

It is, however, possible to write: $C_{ff}(\tau)=\delta(t)*\delta(-t)\big|_{t=\tau}=\delta(\tau)$

b) Let us consider $f(t)=\frac{1}{\Delta}Cr_{\Delta}(t)$

The spectral density modulus is:

$$\left|F(j\omega)\right|=\left|\frac{\sin\frac{\omega\Delta}{2}}{\frac{\omega\Delta}{2}}\right|$$

The spectral density of the energy is written:

$$S(\omega)=\frac{1}{2\pi}\left(\frac{\sin\frac{\omega\Delta}{2}}{\frac{\omega\Delta}{2}}\right)^2$$

The auto-correlation function $C_{ff}(\tau)$ is written:

$$C_{ff}(\tau)=\frac{1}{\Delta}\left(1-\frac{|\tau|}{\Delta}\right)$$

c) Let us consider $f(t) = e^{-\alpha t}u(t)$

The Fourier transform is $f(t) \xleftrightarrow{TF} \dfrac{1}{\alpha + j\omega}$

The spectral density modulus is:

$$\left|F(j\omega)\right| = \frac{1}{\sqrt{\alpha^2 + \omega^2}}$$

The spectral density of the energy is written:

$$S(\omega) = \frac{1}{2\pi}\frac{1}{\alpha^2 + \omega^2}$$

The auto-correlation function $C_{ff}(\tau)$ is written:

$$C_{ff}(\tau) = \frac{1}{2\alpha}e^{-\alpha|\tau|}$$

d) Thus $f(t) = 2\delta(t) + \delta(t-1)$

The Fourier transform is $f(t) \xleftrightarrow{TF} 2 + e^{-j\omega}$

The spectral density modulus is: $\left|F(j\omega)\right| = \sqrt{5 + 4\cos\omega}$

The spectral density of the energy is written:

$$S(\omega) = \frac{1}{2\pi}(5 + 4\cos\omega)$$

The auto-correlation function $C_{ff}(\tau)$ is written:

$$C_{ff}(\tau) = 5\delta(\tau) + 2\delta(\tau+1) + 2\delta(\tau-1)$$

5.3.9. *Exercise 5.9*

The spectral density of energy can be written:

$$S(\omega) = 2\pi\delta(\omega) + 2\pi\left(1-|\omega|\right) \text{ where } |\omega|<1$$

$$S(\omega) = 0 \text{ where } |\omega|>1$$

a) The Fourier transform of the auto-correlation function can be written:

$$C_{ff}(\tau) \xleftrightarrow{TF} 2\pi S(j\omega)$$

We merely determine the inverse Fourier transform of $2\pi S(j\omega)$.

$$\delta_{(\omega)} \overset{TF}{\leftrightarrow} \frac{1}{2\pi} \quad \delta(\omega) \xleftrightarrow{TF} \frac{1}{2\pi}$$

$$1-|\omega| \overset{TF}{\leftrightarrow} \frac{1}{2\pi} 2\int_0^1 (1-\omega)\cos\tau\omega d\omega$$

Hence $1-|\omega| \overset{TF}{\leftrightarrow} \frac{1}{2\pi}\left(\frac{\sin \tau/2}{\tau/2}\right)^2$

After calculating: $C_{ff}(\tau) = 2\pi\left(1+\left(\frac{\sin \tau/2}{\tau/2}\right)^2\right)$

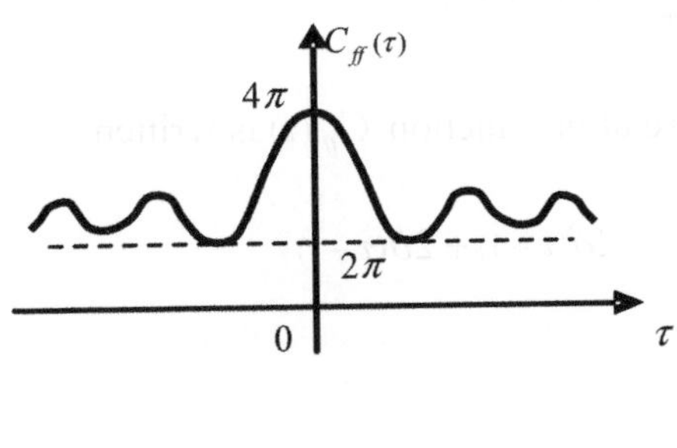

b) We consider a signal $f(t)$ written as follows:

$$f(t) = a_0 + \phi(t)$$

a_0 is the DC component

$f(t) = a_0 + \phi(t)$ with $\phi(t)$ alternating (AC) component such that: $\int_{-\infty}^{+\infty} \phi(t)dt = 0$

The auto-correlation function can appear under the following form:

$$C_{ff}(\tau) = a_0^2 + C_{\phi\phi}(\tau)$$

Also: $C_{\phi\phi}(\tau) = 2\pi + 2\pi\left(\frac{\sin \tau/2}{\tau/2}\right)^2$

Thus, by identifying the mean direct (DC) energy:

$$\boxed{a_0^2 = 2\pi}$$

c) The variable component of the auto-correlation function is written as follows:

$$C_{\phi\phi}(\tau) = 2\pi\left(\frac{\sin \tau/2}{\tau/2}\right)^2$$

The mean alternating AC energy can then be written: $C_{\phi\phi}(0) = 2\pi$

5.3.10. *Exercise 5.10*

a) If the filter is matched, we can write the following:

$$h(t) \xleftrightarrow{TF} H(j\omega) = X^*(j\omega)$$

Knowing that:

$$x(t) \xleftrightarrow{TF} X(j\omega),$$

we get: $x^*_{(-t)} \overset{TF}{\leftrightarrow} X^*_{(j\omega)}$ $x^*(-t) \xleftrightarrow{TF} X^*(j\omega)$

From this, we establish that:

$$h(t) = x^*(-t)$$

If the input function is real:

$$h(t) = x(-t)$$

b) We write for the output value:

$$y(t) = h(t) * x(t)$$

If the filter is matched:

$$h(t) = x^*(-t)$$

Thus after replacement:

$$y(t) = x^*(-t) * x(t) = \int_{-\infty}^{+\infty} x(\tau)x^*(\tau - t)d\tau$$

Or:

$$y(t) = C_{xx}(t)$$

5.3.11 *Exercise 5.11*

Let $y(t)$ be the circuit response. If it corresponds to the cross-correlation between functions $y(t)$ and $g(t) = e^{at}u(-t)$, it is possible to write:

$$y(t) = f(t) * g(-t)$$

Thus by replacing $g(t)$:

$$y(t) = f(t) * \left(e^{-at}u(t)\right)$$

After identification we obtain:

$$h(t) = e^{-at}u(t)$$

The transfer function is the Laplace transform of $g(t)$:

$$H(p) = \frac{1}{p+a}$$

If:

$$a = 1 \; H(p) = \frac{1}{p+1}$$

It is the transfer function of an RC cell:

$$H(p) = \frac{1}{RCp+1} \text{ with } RC = 1s$$

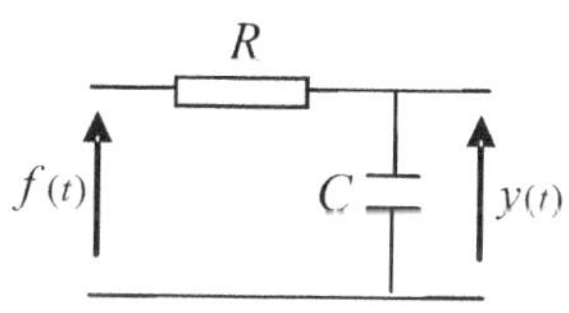

5.3.12 *Exercise 5.12*

If $f_1(t)$ and $f_2 g(t)$ are two complex time-functions, the cross-correlation function is written:

$$C_{12}(\tau) = \int_{-\infty}^{+\infty} f_1(t) f_2^*(t-\tau)dt = f_1(\tau) * f_2^*(-\tau)$$

If $f_1(\tau) \xleftrightarrow{TF} F_1(j\omega)$ and $f_1(\tau) \xleftrightarrow{TF} F_2(j\omega)$

We know that:

$$f_2^*(-\tau) \xleftrightarrow{TF} F_2^*(j\omega)$$

In this case:

$$C_{12}(\tau) \xleftrightarrow{TF} F_1(j\omega) F_2^*(j\omega)$$

In the same way, we can state the following:

$$C_{21}(\tau)=\int_{-\infty}^{+\infty} f_2(t)f_1^*(t-\tau)dt = f_2(\tau) * f_1^*(-\tau)$$

$$\boxed{C_{21}(\tau)\xleftrightarrow{TF} F_2(j\omega)F_1^*(j\omega)}$$

Using the inverse Fourier transform of $C_{12}(\tau)$, we write:

$$C_{12}(\tau)=\frac{1}{2\pi}\int_{-\infty}^{+\infty} F_1(j\omega)F_2^*(j\omega)e^{j\omega\tau}d\omega$$

Also: $C_{12}(\tau)=\int_{-\infty}^{+\infty} f_1(t)f_2^*(t-\tau)dt$

$\tau=0$ is a particular case and allows us to write:

$$\boxed{\int_{-\infty}^{+\infty} f_1(t)f_2^*(t)dt=\frac{1}{2\pi}\int_{-\infty}^{+\infty} F_1(j\omega)F_2^*(j\omega)d\omega}$$

The same type of calculation allows us to write:

$$\boxed{\int_{-\infty}^{+\infty} f_1(t)f_2^*(-t)dt=\frac{1}{2\pi}\int_{-\infty}^{+\infty} F_1(j\omega)F_2^*(-j\omega)d\omega}$$

5.3.13. *Exercise 5.13*

System response $f_2(t)$ of the linear system is written:

$$f_2(t)=h(t) * f_1(t)$$

The cross-correlation function can take the form of a convolution.

$$C_{21}(\tau)=f_2(\tau) * f_1^*(-\tau)$$

Hence: $C_{21}(\tau)=h(\tau) * f_1(\tau) * f_1^*(\tau)$

Such as $C_{11}(\tau)=f_1(\tau)*f_1^*(-\tau)$

We obtain: $\boxed{C_{21}(\tau)=h(\tau)*C_{11}(\tau)}$

As such: $C_{22}(\tau)=f_2(\tau)*f_2^*(-\tau)$

or

$$C_{22}(\tau)=\left(f_1(\tau)*h(\tau)\right)*\left(f_1^*(-\tau)*h^*(-\tau)\right)=f_1(\tau)*f_1^*(-\tau)*h(\tau)*h^*(-\tau)$$

Hence: $C_{22}(\tau)=C_{11}(\tau)*C_{hh}(\tau)$

This can be written as follows:

$$\boxed{C_{22}(\tau)=\int_{-\infty}^{+\infty}C_{hh}(t)*C_{11}(\tau-t)dt}$$

The Fourier transform of $C_{22}(\tau)$ can be written:

$$TF\left(C_{22}(\tau)\right)=TF\left(C_{11}(\tau)\right)TF\left(C_{hh}(\tau)\right)$$

The Fourier transform of the auto-correlation function is equal to the square of the modulus of the Fourier transform:

Hence: $\boxed{TF\left(C_{22}(\tau)\right)=\left|H(j\omega)\right|^2 TF\left(C_{11}(\tau)\right)}$

5.3.14. *Exercise 5.14*

a) The energy of the signal is written:

$$E=\int_{-\infty}^{+\infty}f^2(t)dt$$

$\forall a$ and $b \in R^*$, the integral is infinite. The signal therefore does not have a finite energy.

This signal can be considered periodic if $\frac{a}{b}$ is a rational number.

b) Signal $f(t)$ can appear under the following form:

$$f(t) = f_1(t) + f_2(t) \text{ with } f_1(t) = \sin at \text{ and } f_2(t) = \sin bt$$

In this case $C_{ff}(\tau) = C_{f_1f_1}(\tau) + C_{f_1f_2}(\tau) + C_{f_2f_1}(\tau) + C_{f_2f_2}(\tau)$

$$C_{f_1f_1}(\tau) = \lim_{T\to\infty} \frac{1}{T}\int_0^T \sin at \sin a(t-\tau)dt$$

or $C_{f_1f_1}(\tau) = \frac{1}{2}\cos a\tau$

and $C_{f_2f_2}(\tau) = \frac{1}{2}\cos b\tau$

$$C_{f_1f_2}(\tau) = \lim_{T\to\infty} \frac{1}{T}\int_0^T \sin at \sin b(t-\tau)dt$$

$$= \lim_{T\to\infty} \frac{\cos b\tau}{2}\left(\frac{\sin(a-b)\frac{T}{2}}{(a-b)\frac{T}{2}} - \frac{\sin(a+b)\frac{T}{2}}{(a+b)\frac{T}{2}}\right)$$

or $C_{f_1f_2}(\tau) = 0$

and $C_{f_2f_1}(\tau) = 0$

Hence: $\boxed{C_{ff}(\tau) = \frac{1}{2}(\cos a\tau + \cos b\tau)}$

The limit of $C_{ff}(\tau)$ is not null when $\tau \to \infty$, as is the case for signals with finite energy.

c) For $a = 2$ and $b = 5$:

$$C_{ff}(\tau) = \frac{1}{2}(\cos 2\tau + \cos 5\tau)$$

This function is periodic with a period of π

For $a = 1$ and $b = \sqrt{2}$:

$$C_{ff}(\tau) = \frac{1}{2}\left(\cos \tau + \cos \sqrt{2}\tau\right)$$

This function is not periodic.

6

Signal Sampling

6.1. Theoretical background

6.1.1. *Sampling principle*

In many cases, it is pointless to process continuous (analog) signals. We will see that knowing the value of the signal at specific points is sufficient.

These points are called the sampling instants.

The values of the signal at these points are called the samples.

Let us consider function $f(t)$ and multiply this function by a Dirac impulse centered at instant $t-t_0$.

$$f(t)\delta(t-t_0)=f(t_0)\delta(t-t_0)$$

This mathematical operation represents the sampling principle of a signal.

A couple of points on this notation $\delta(t)$:

– We could define this type of function as:

$$\delta(t) = \begin{cases} 0 & t \neq 0 \\ \infty & t = 0 \end{cases} \quad \text{and} \quad \int_{-\varepsilon}^{+\varepsilon} \delta(t)dt = 1 \quad \forall \varepsilon > 0 \text{ no matter how small}$$

But an ordinary function null everywhere except at one point will have an integral said to be null according to Riemann This means $\delta(t)$ cannot be an "ordinary" function and we therefore define it as:

$$\int_{-\infty}^{+\infty} \varphi(t)\delta(t)dt = \varphi(0)$$

where $\varphi(t)$ is an infinitely derivable function (C^{∞}) at $t = 0$ and is referred to as a test function [ROD 93, DEM 96, LES 12].

– An alternative definition of $\delta(t)$ can be provided by:

$$\int_{a}^{b} \phi(t)\delta(t)dt = a < 0 < b \begin{cases} \phi(0) & a < 0 < b \\ 0 & a < b < 0 \text{ or } 0 < a < b \\ \text{undefined} & a = 0 \text{ or } b = 0 \end{cases}$$

However these last 2 integral notations are not to be considered as integrals according to Riemann. Also, this is why $\delta(t)$ is referred to as a distribution rather than a function. As $\varphi(t)$ is a function that is indefinitely derivable in all points, we can write:

$$\int_{-\infty}^{+\infty} \phi(t)\delta(t - t_0)dt = \phi(t_0)$$

6.1.2. *Ideal sampling*

If we multiply a signal $f(t)$ by a Dirac comb function $\delta_T(t)$ represented in Figure 6.1, the sample signal can be written:

$$\boxed{f_E(t) = f(t)\delta_T(t)} \quad [6.1]$$

It appears as a series of Dirac impulses located at integer multiples of T and has a weight equal to the value of the function at the sampling moment.

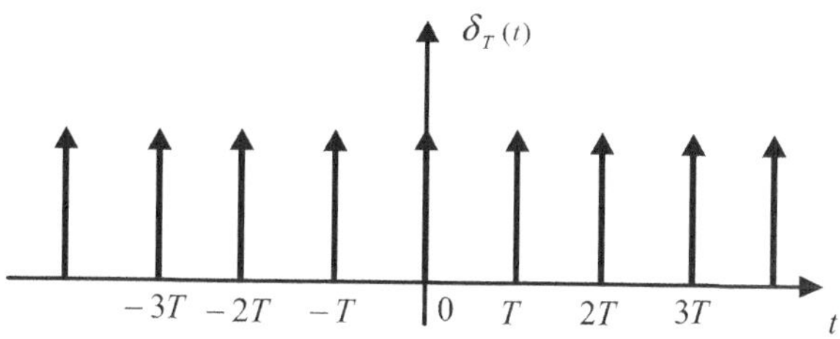

Figure 6.1. *Dirac comb function*

An example of this is represented in Figure 6.2.

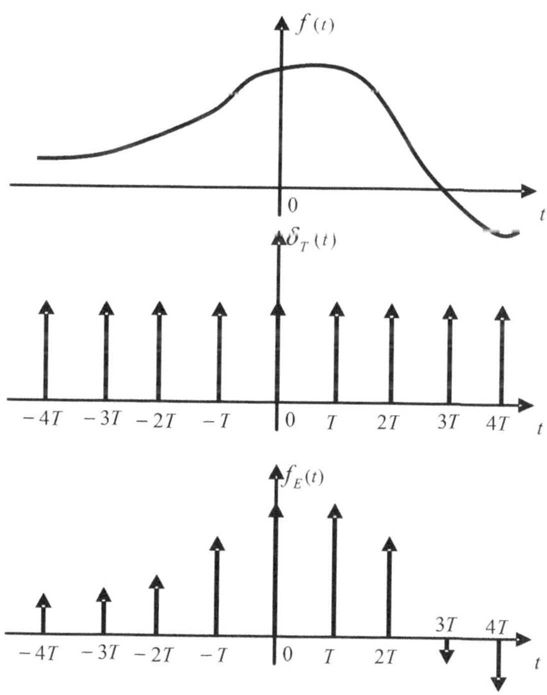

Figure 6.2. *Sampling principle*

The mathematical expression of the Dirac comb function is:

$$\delta_T = \sum_{(t)n=-\infty}^{n=+\infty} \delta(t - nT)$$

The sampled signal is then written:

$$f_E(t) = f(t) \sum_{n=-\infty}^{n=+\infty} \delta(t - nT)$$

$$\text{hence: } \boxed{f_E(t) = \sum_{n=-\infty}^{n=+\infty} f(nT)\delta(t - nT)} \qquad [6.2]$$

The Fourier transform of the Dirac comb function is:

$$\delta_T(t) \xleftrightarrow{TF} \frac{2\pi}{T} \sum_{k=-\infty}^{k=+\infty} \delta(\omega - k\omega_0) = \omega_0 \delta_{\omega_0}(\omega) \text{ where } \omega_0 = \frac{2\pi}{T}$$

Consider the following: $f(t) \xleftrightarrow{TF} F(j\omega)$.

The Fourier transform of the sampled signal can then be written:

$$F_E(j\omega) = \frac{1}{2\pi}\left(F(j\omega) * \omega_0 \delta_{\omega_0}(\omega)\right)$$

or:

$$\boxed{F_E(j\omega) = \frac{1}{T} \sum_{n=-\infty}^{n=+\infty} F(\omega - n\omega_0)} \qquad [6.3]$$

The spectrum of the sampled signal corresponds to the one repeated on the frequency axis with a period of ω_0. Figure 6.3 presents two examples of this.

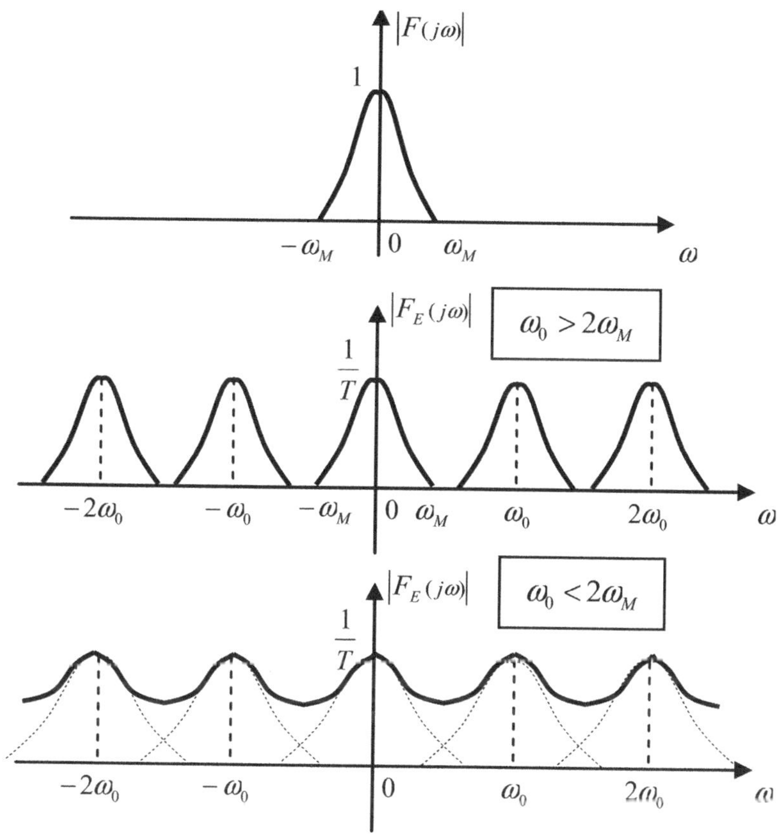

Figure 6.3. *Spectrum of the sampled signal*

Note that if $\omega_0 > 2\omega_M$, the signal is not distorted.

If we multiply this spectrum with a rectangular function $Cr_{\omega_0}(\omega)$ which has a width of ω_0, we end up with $F_{(j\omega)}$, the initial spectrum (see Figure 6.4).

This is Shannon's theorem.

$$\boxed{TF_E(j\omega)Cr_{\omega_0}(\omega) = F(j\omega)} \quad [6.4]$$

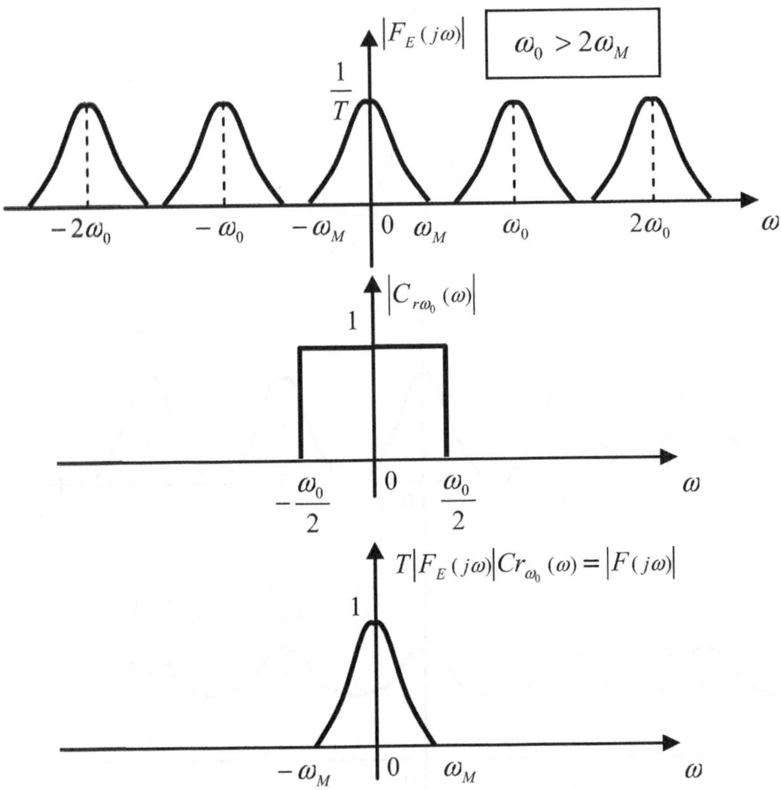

Figure 6.4. *Re-encountering the initial signal*

In practice the sample pulses are not infinitely narrow. The following paragraph will allow us to study how the width of the pulse affects the previous results.

6.1.3. *Finite width sampling*

To make the comparison between results easier, we will select surface unit pulses such as the one shown in Figure 6.5.

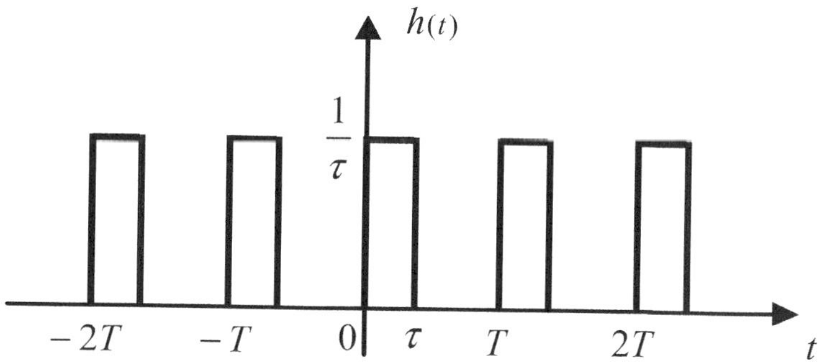

Figure 6.5. *Sample pulses*

These sample pulses can be noted as follows:

$$h(t) = \frac{1}{\tau} \sum_{n=-\infty}^{n=+\infty} Cr_\tau (t - \frac{\tau}{2} - nT)$$

or:

$$h(t) = \frac{1}{\tau} \sum_{n=-\infty}^{n=+\infty} Cr_\tau (t - \frac{\tau}{2}) * \delta(t - nT) \qquad [6.5]$$

For the sampled signal we write:

$$f_E(t) - f(t)h(t)$$

Thus: $f_E(t) = \frac{1}{\tau} \sum_{n=-\infty}^{n=+\infty} f(t) Cr_\tau (t - \frac{\tau}{2}) * \delta(t - nT)$

Let us now calculate the Fourier transforms: $\begin{cases} f_E(t) \xleftrightarrow{TF} F_E(j\omega) \\ f(t) \xleftrightarrow{TF} F(j\omega) \\ h(t) \xleftrightarrow{TF} H(j\omega) \end{cases}$

$H(j\omega)$ can be easily calculated:

$$H(j\omega) = \frac{1}{\tau}\sum_{n=-\infty}^{n=+\infty} \tau \frac{\sin\frac{\omega\tau}{2}}{\frac{\omega\tau}{2}} e^{-j\frac{\omega\tau}{2}} \omega_0 \delta(\omega - n\omega_0) \text{ where } \omega_0 = \frac{2\pi}{T}$$

This can be noted under the following form:

$$\boxed{H(j\omega) = \omega_0 \sum_{n=-\infty}^{n=+\infty} \frac{\sin\frac{n\omega_0\tau}{2}}{\frac{n\omega_0\tau}{2}} e^{-j\frac{n\omega_0\tau}{2}} \delta(\omega - n\omega_0)} \qquad [6.6]$$

We can then determine $F_{E(j\omega)}$:

$$F_E(j\omega) = \frac{1}{2\pi} F(j\omega) * H(j\omega)$$

$$F_E(j\omega) = \frac{1}{T}\sum_{n=-\infty}^{n=+\infty} \frac{\sin\frac{n\omega_0\tau}{2}}{\frac{n\omega_0\tau}{2}} e^{-j\frac{n\omega_0\tau}{2}} F(j\omega) * \delta(\omega - n\omega_0)$$

Then: $$\boxed{F_E(j\omega) = \frac{1}{T}\sum_{n=-\infty}^{n=+\infty} \frac{\sin\frac{n\omega_0\tau}{2}}{\frac{n\omega_0\tau}{2}} e^{-j\frac{n\omega_0\tau}{2}} F(\omega - n\omega_0)} \qquad [6.7]$$

Let us note that the sine cardinal only represents a change in amplitude for frequencies which are multiples of ω_0.

If $n = 0$, there are no changes.

Let us present the graphical representation of this phenomenon in Figure 6.6.

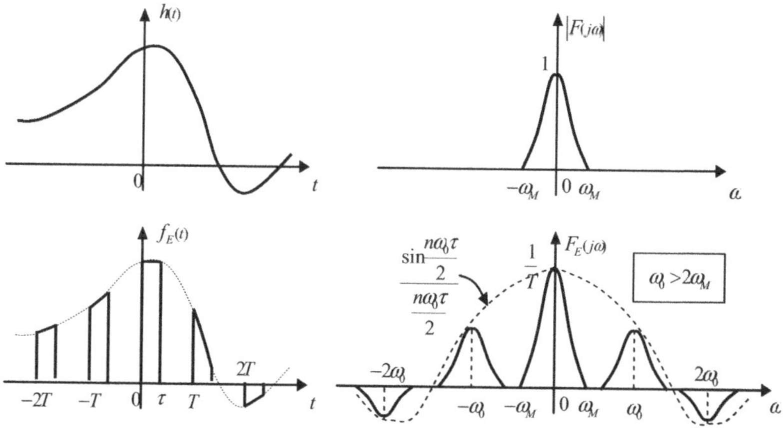

Figure 6.6. *Finite width sampling*

In the case where $\omega_0 > 2\omega_M$ if we multiply $F_E(j\omega)$ by $Cr_{\omega_0}(\omega)$, we obtain:

$$\boxed{TF_E(j\omega)Cr_{\omega_0}(\omega) = F(j\omega)} \quad [6.8]$$

This result is the same as the previous.

6.1.4. *Sample and hold (S/H) sampling*

In this case, function $f(t)$ is no longer allowed evolve during the pulse. Its value is fixed at the start of the pulse. This is represented in Figure 6.7.

In practice this approach is the most utilized as it maintains constant the value of the sample for the duration of an interval allowing it to be converted from analog to digital.

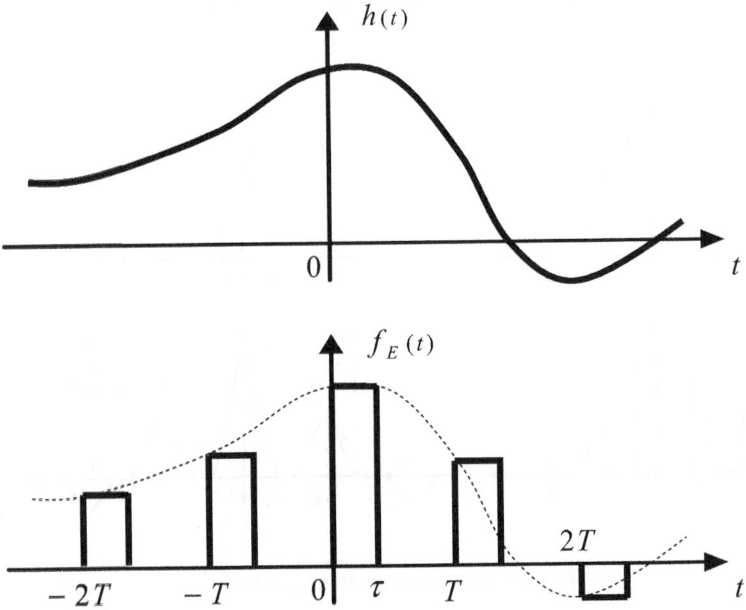

Figure 6.7. *Sample and hold (S/H) sampling*

The sampled signal is then written:

$$f_E(t) = \frac{1}{\tau} \sum_{n=-\infty}^{n=+\infty} f(nT) Cr_\tau(t - \frac{\tau}{2} - nT)$$

We can note this the same way as previously:

$$f_E(t) = \frac{1}{\tau} \sum_{n=-\infty}^{n=+\infty} f(nT) Cr_\tau(t - \frac{\tau}{2}) * \delta(t - nT)$$

$$f_E(t) = \frac{1}{\tau} Cr_\tau(t - \frac{\tau}{2}) * \sum_{n=-\infty}^{n=+\infty} f(nT) \delta(t - nT)$$

The second member of the convolution presents an ideal-case expression of the sampled signal.

This allows us to establish the Fourier transform.

$$F_E(j\omega) = \frac{\sin\frac{\omega\tau}{2}}{\frac{\omega\tau}{2}} e^{-j\frac{\omega\tau}{2}} \frac{1}{T} \sum_{n=-\infty}^{n=+\infty} F(\omega - n\omega_0) \qquad [6.9]$$

The spectrum of the signal sampled in an ideal case is therefore multiplied by $\frac{\sin\frac{\omega\tau}{2}}{\frac{\omega\tau}{2}} e^{-j\frac{\omega\tau}{2}}$ which represents a modulation of the modulus of the spectrum by $\frac{\sin x}{x}$ and a additional phase shift $-\frac{\omega\tau}{2}$.

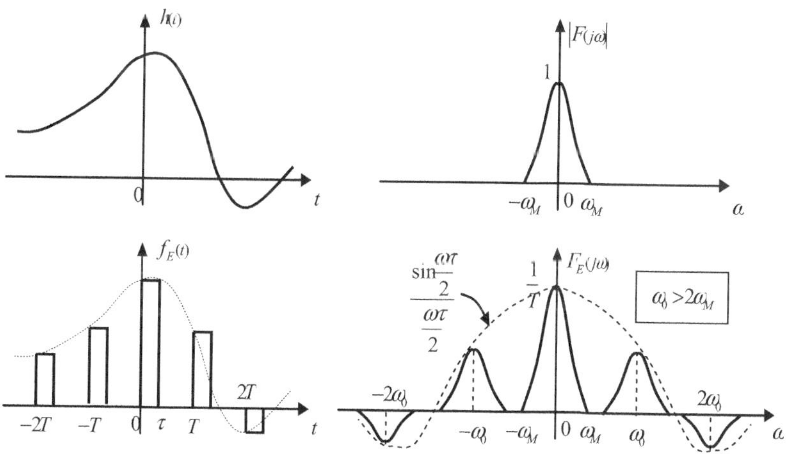

In the case where $\omega_0 > 2\omega_M$, if we multiply $F_E(j\omega)$ by $Cr_{\omega_0(\omega)}$, we obtain:

$$\boxed{TF_E(j\omega) = \frac{\sin\frac{\omega\tau}{2}}{\frac{\omega\tau}{2}} e^{-j\frac{\omega\tau}{2}} F(j\omega)} \qquad [6.10]$$

After filtering, this results in changes to the modulus and the phase of the spectrum.

Let us study how T and τ affect the error induced by the sample and hold (S/H) sampler.

Let $\tau = \lambda T$ with $\lambda < 1$

And $\omega_0 = 2\alpha\omega_M$ with $\alpha > 1$

The error-inducing factor can therefore be noted:

$$\frac{\sin\frac{\pi}{2}\frac{\lambda}{\alpha}\frac{\omega}{\omega_M}}{\frac{\pi}{2}\frac{\lambda}{\alpha}\frac{\omega}{\omega_M}}$$

If we wish to limit the error to *1%* until ω_M, the following must be true:

$$\frac{\sin\frac{\pi}{2}\frac{\lambda}{\alpha}}{\frac{\pi}{2}\frac{\lambda}{\alpha}} \geq 0.99 \quad \text{meaning} \quad \frac{\lambda}{\alpha} \leq 0.16$$

If we wish to limit the error to 0.1%, the limit value becomes $\frac{\lambda}{\alpha} \leq 0.012$.

For the phase change:

$$\Delta\varphi = \frac{\pi}{2}\frac{\lambda}{\alpha}\frac{\omega}{\omega_M}$$

If we limit phase error to under 5°, we obtain $\dfrac{\lambda}{\alpha} \leq 0.055$

6.2. Exercises

The exercises presented in this chapter come from B.P. Lathi [LAT 66].

6.2.1. *Exercise 6.1*

Determine the minimum sampling frequency for the following signals:

$$x_1(t) = \frac{\sin 100t}{100t} \quad \text{and} \quad x_2(t) = \left(\frac{\sin 100t}{100t}\right)^2$$

6.2.2. *Exercise 6.2*

Let us consider the following signals. Their spectral occupation is not finite.

$$f_1(t) = e^{-5|t|},\ f_2(t) = te^{-5t}u_{(t)},\ f_3(t) = Cr_{10}(t) \text{ and } f_4(t) = e^{-t}\sin 100t\, u(t)$$

Suggest a criterion that would allow us to consider these signals as spectrally limited and from that, deduce the minimum sampling frequency which would allow us to reproduce the function with minimal distortion.

6.2.3. *Exercise 6.3*

Plot the spectrum of $f(t)=\dfrac{\sin 200\pi t}{200\pi t}$. This signal is sampled by $\delta_T(t)$. Plot the spectrum of the sampled signal for three values from the sampling period:

$$T=\frac{1}{400}s,\ T=\frac{1}{200}s,\text{ and } T=\frac{1}{100}s.$$

6.2.4. *Exercise 6.4*

An ideal low-pass filter for transfer function: $H(j\omega)=H_0e^{-j\omega t_0}Cr_{2\omega_0}(\omega)$ is used as an interpolation filter for the following sampled signal:

$$x_E(t)=Tx(t)\delta_T(t) \text{ where } T<\frac{\pi}{\omega_0}$$

Signal $x(t)$ has a limited band, meaning $X(j\omega)=0$ for $|\omega|>\omega_0$.

Demonstrate that the response from the filter is given by:

$$y(t)=H_0x(t-t_0)$$

6.2.5. *Exercise 6.5*

Consider a signal $f(t)$ that exists only on a finite time-interval:

$$f(t)=0 \text{ for } |t|>T$$

Demonstrate that Fourier transform $F(j\omega)$ is entirely determined by the knowledge we have of $F(j\omega)$ through sampling $\dfrac{\pi}{T}$ at distant frequencies.

We then only need to know $F_n(n\frac{\pi}{T})$.

Demonstrate that $F(j\omega)$ can be written under the following form:

$$F(j\omega)=\sum_{n=-\infty}^{n=+\infty}F_n(n\frac{\pi}{T})\frac{\sin(\omega T-n\pi)}{\omega T-n\pi}$$

6.2.6. *Exercise 6.6*

Let $f(t)$ be a function with a spectrum that is limited to frequency ω_M and $f_E(t)$ the signal sampled at a frequency superior to $2\omega_M$. $f(t)$ can then be reproduced after passing through an ideal low-pass filter.

Consider the following circuit:

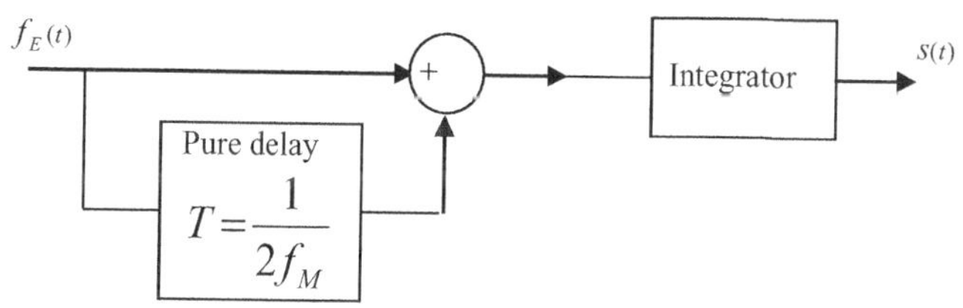

a) What is the transfer function of this circuit?

b) Determine the relation between Fourier transforms of $s(t)$ and $f(t)$.

c) Demonstrate that $s(t)$ is not very different from $f(t)$.

d) Compare with response from the first order low-pass filter.

6.2.7. *Exercise 6.7*

Generalize the sampling theorem to the case where $f(t)$ is a band-pass signal, meaning the spectrum is limited to interval $|\omega|\in[\omega_m,\omega_M]$.

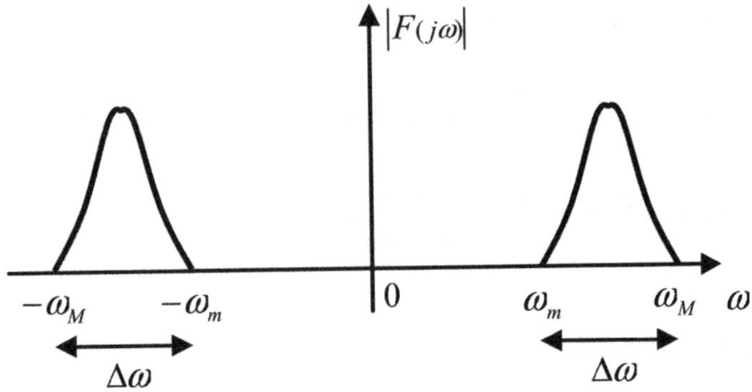

Demonstrate that it is possible to sample frequencies under $2\omega_M$:

a) Considering the following example:

$$\omega_M = 2.5\Delta\omega\,,\ \Delta\omega = 2 \text{ and } \omega_{\text{sample}} = 1.1\omega_M$$

b) Considering a generic case.

6.2.8. *Exercice 6.8*

Let $x(t)$ be a periodic function of period $T = \dfrac{2\pi}{\omega_0}$ which has no spectral components of pulsation $\omega > N\omega_0$.

a) Demonstrate that $x(t)$ is entirely specified by the $2N+1$ samples $x(k\frac{T}{2N+1})$ and that it is given by the following equation:

$$x(t) = \frac{1}{2\pi}\sum_{k=-N}^{k=+N} x(k\frac{T}{2N+1})\frac{\sin\left((2N+1)\frac{\pi}{T}\left(t-k\frac{T}{2N+1}\right)\right)}{\sin\left(\frac{\pi}{T}\left(t-k\frac{T}{2N+1}\right)\right)}$$

6.3. Solutions to the exercises

6.3.1. *Exercise 6.1*

$$x_1(t) = \frac{\sin 100t}{100t} \overset{TF}{\leftrightarrow} X_1(t) = \frac{2\pi}{200} Cr_{200}(j\omega)$$

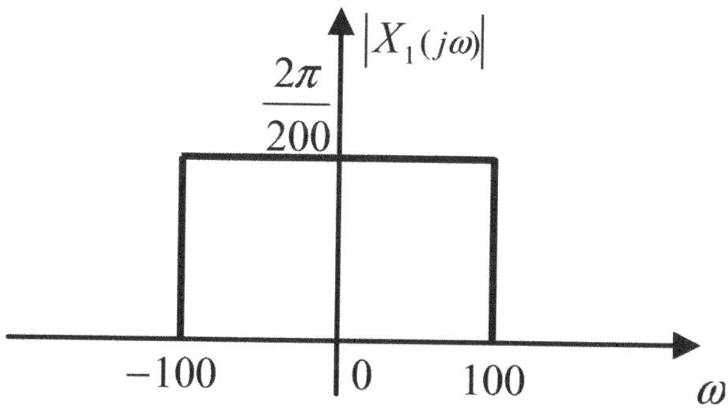

The maximum pulsation contained in the spectrum of $x_1(t)$ is $\omega_M = 100rds^{-1}$.

This indicates that: $\omega_E \geq 200\,rds^{-1}$

As such:

$$x_2(t) = \left(\frac{\sin 100t}{100t}\right)^2 \xleftrightarrow{TF} X_2(j\omega) = \frac{\pi}{200}\left(1 - \frac{|\omega|}{200}\right)$$

Hence $\omega_M = 200rds^{-1}$ and $\omega_E \geq 400\,rds^{-1}$

6.3.2. *Exercise 6.2*

These different signals occupy an infinite frequency band. The sampling theorem requires for the signal spectrum to have a limited band, we must therefore estimate a pulse ω_M from which the amplitude of the spectrum will be considered to be null.

Our criterion will be a maximum frequency such that the amplitude of the spectrum will be under100 times its maximum value.

$$\omega_M \Rightarrow \left|F(j\omega_M)\right| < \frac{\left|F(j\omega)\right|_{max}}{100}$$

– $f_1(t) = e^{-5|t|} \xleftrightarrow{TF} F_1(j\omega) = \dfrac{10}{25+\omega^2}$

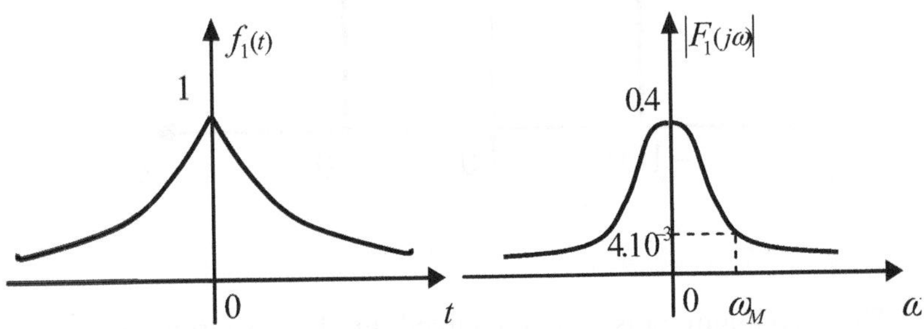

We deduce that $\left|F_1(j\omega)\right|_{max} = 0.4$ where $\omega = 0$.

We deduce that: $\boxed{\omega_M \Rightarrow \dfrac{10}{25+\omega_M^2} \leq 4.10^{-3}}$

or: $\omega_M = 50rds^{-1}$

We will select a sampling frequency such that:

$\boxed{\omega_E = 100rds^{-1}}$

– $f_2(t) = te^{-5t}u(t) \xleftrightarrow{TF} F_2(j\omega) = \dfrac{1}{(2+j\omega)^2}$

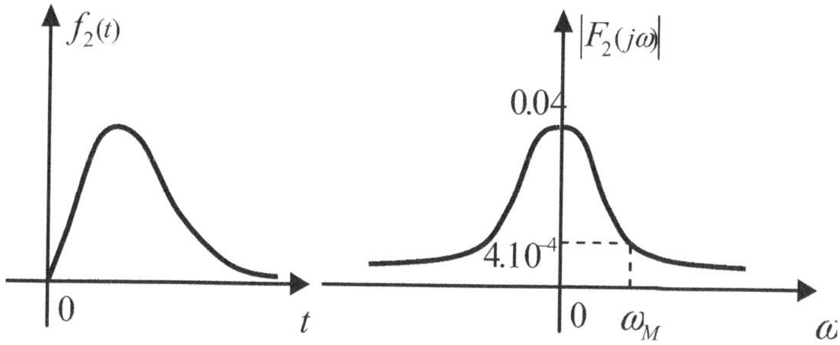

From this we establish that $|F_2(j\omega)|_{max} = 0.04$ for $\omega = 0$.

This indicates that: $\boxed{\omega_M \Rightarrow \dfrac{1}{25+\omega_M^2} \leq 4.10^{-4}}$

Thus: $\omega_M = 50 rds^{-1}$

We select a sampling frequency such that:

$\boxed{\omega_E = 100 rds^{-1}}$

We note that the argument of $F_2(j\omega)$ is equal to -180° for an infinite frequency and 178° for $\omega = \omega_M = 50\ rds^{-1}$, thus an error of 2°.

– $f_3(t) = Cr_{10}(t) \xleftrightarrow{TF} F_3(j\omega) = 10\dfrac{\sin 5\omega}{5\omega}$

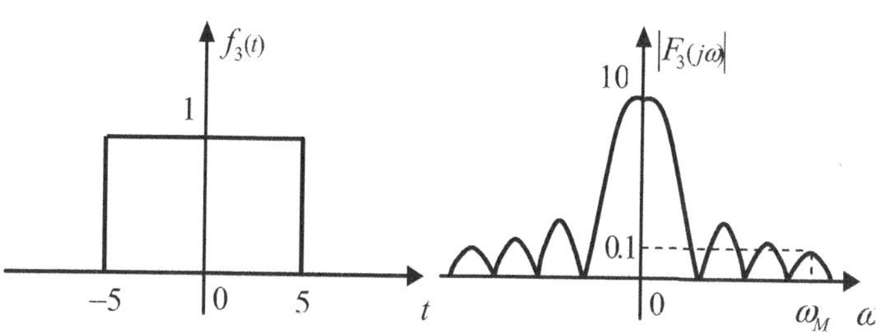

From this we establish that $|F_3(j\omega)|_{\max} = 10$ for $\omega = 0$.

This indicates that: $\boxed{\omega_M \Rightarrow \left|\dfrac{\sin 5\omega_M}{5\omega_M}\right| \leq 10^{-2}}$

The function is not monotonic – there are actually an endless number of solutions to this equation.

We choose to solve this equation with the value of the pulse corresponding to the maximum of $\left|\dfrac{\sin x}{x}\right|$ immediately under the limit value. This value can be obtained using the tables.

Thus: $\omega_M \approx 20 rds^{-1}$

We select a sampling frequency such that:

$\boxed{\omega_E \approx 40 rds^{-1}}$

– $f_4(t) = e^{-t}\sin 100t u(t) \xleftrightarrow{TF} F_4(j\omega) = \dfrac{100}{(1+j\omega)^2 + 100^2}$

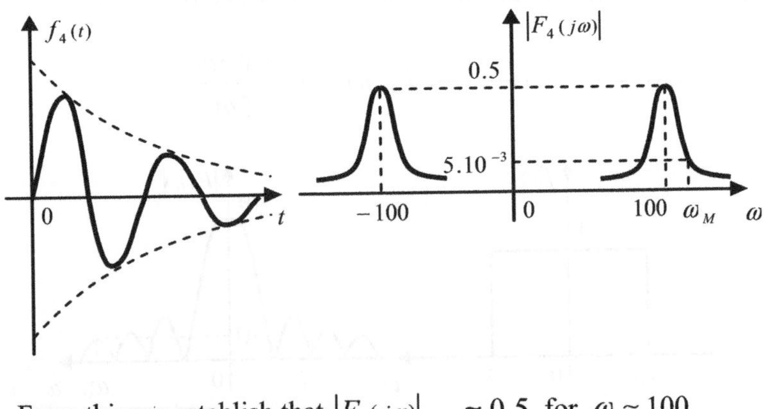

From this we establish that $|F_4(j\omega)|_{\max} \approx 0.5$ for $\omega \approx 100$.

This indicates that:

$$\boxed{\omega_M \Rightarrow \frac{100}{\sqrt{\left(100^2 - \omega_M^2\right)^2 + 4\omega_M^2}} \leq 510^{-3}}$$

Thus: $\omega_M \approx \sqrt{3}10^2 rds^{-1}$

We select a sampling frequency such that:

$$\boxed{\omega_E \approx 2\sqrt{3}10^2 rds^{-1}}$$

6.3.3. *Exercise 6.3*

Let us plot the spectrum of $f(t)$:

$$f(t) = \frac{\sin 200\pi t}{200\pi t} \xleftrightarrow{TF} F(j\omega) = \frac{1}{200} C_{r400\pi(t)}$$

The spectrum of the signal sampled by a Dirac comb function is easily found by repeating it on the frequency axis of the signal spectrum.

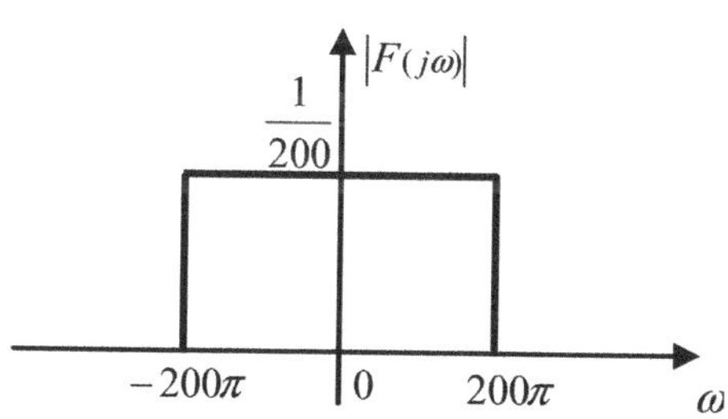

a) $T = \dfrac{1}{400}$

The sampling pulse is equal to $\omega_E = 800\pi$

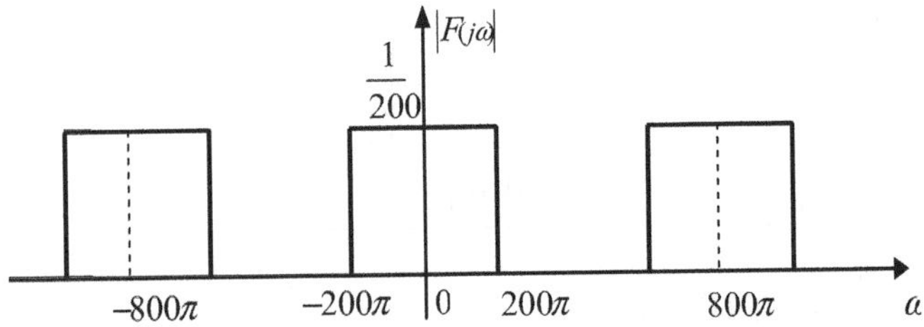

The sampling frequency is superior to Shannon's frequency.

Reconstructing this signal through filtering is not difficult.

b) $T = \dfrac{1}{200}$

The sampling pulse is equal to $\omega_E = 400\pi$

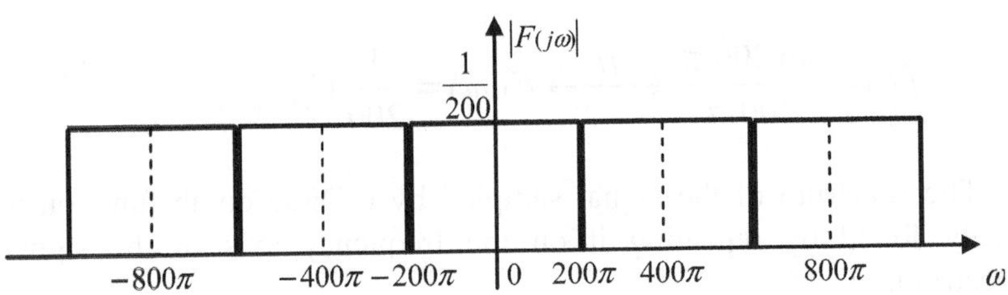

The sampling frequency is equal to the Shannon's frequency.

Reconstructing this signal through filtering is difficult if not altogether impossible in practice due to it being impossible to design an ideal filter.

c) $T = \dfrac{1}{100}$

The sampling pulse is equal to $\omega_E = 200\pi$

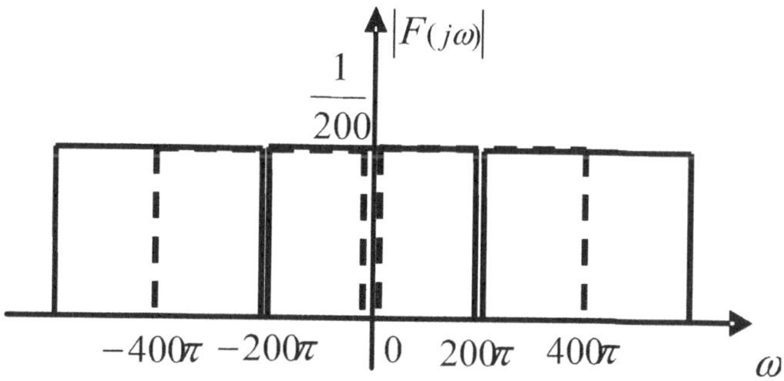

The sampling frequency is lower than Shannon's frequency.

Reconstructing this signal through filtering is impossible due to the spectra overlapping.

6.3.4. *Exercice 6.4*

If the spectrum of $x(t)$ has bounded support and if the sampling frequency is higher than Shannon's frequency, the spectrum of the sampled frequency can be written as follows:

$$X_E(j\omega) = \sum_{n=-\infty}^{n=+\infty} X(\omega - n\omega_E)$$

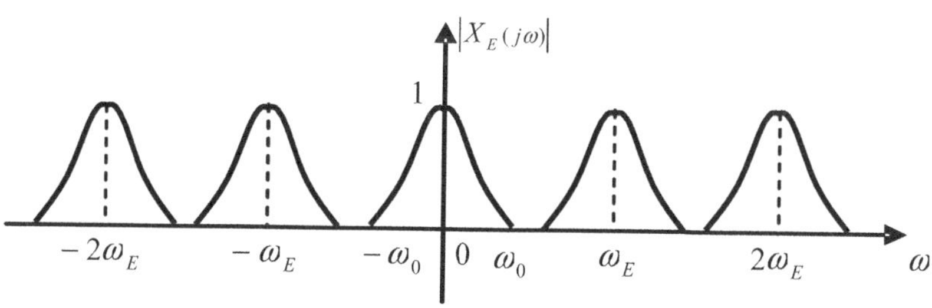

At the output of the filter, we can write:

$$Y(j\omega) = H(j\omega) * X_E(j\omega)$$

Thus: $$Y(j\omega) = H_0 e^{-j\omega t_0} Cr_{2\omega_0}(\omega) \sum_{n=-\infty}^{n=+\infty} X(\omega - n\omega_E)$$

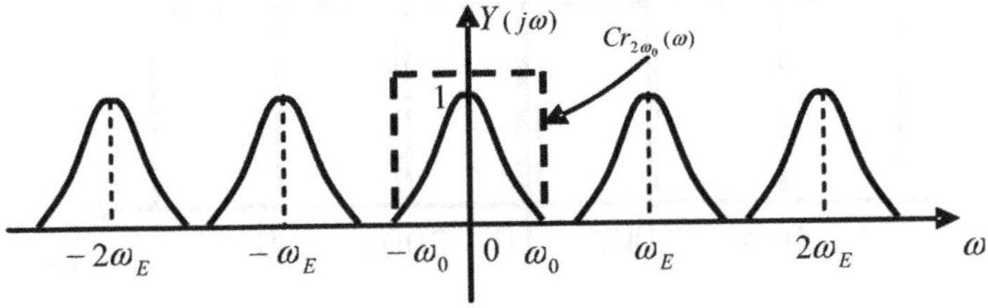

This can be written: $Y(j\omega) = H_0 e^{-j\omega t_0} X(j\omega)$.

Using inverse Fourier transform, we obtain:

$$\boxed{y(t) = H_0 x(t - t_0)}$$

6.3.5. *Exercise 6.5*

Let us express spectrum $F(j\omega)$ sampled by a Dirac comb function of period $\omega_E = \frac{\pi}{T}$.

We obtain a series of samples that can be written as follows:

$$F_E(j\omega) = F(j\omega) \sum_{n=-\infty}^{n=+\infty} \delta(\omega - n\frac{\pi}{T})$$

$$= \sum_{n=-\infty}^{n=+\infty} F(n\frac{\pi}{T}) \delta(\omega - n\frac{\pi}{T})$$

From this, we can establish the inverse Fourier transform:

$$f_E(t) = f(t) * \frac{T}{\pi} \sum_{n=-\infty}^{n=+\infty} \delta(t - 2nT)$$

Sampling $F(j\omega)$, we obtain through inverse transform a periodic function $f_E(t)$. The base function of $f_E(t)$ is $f(t)$.

We can then write:

$$f(t) = f_E(t)\frac{\pi}{T}Cr_{2T}(t)$$

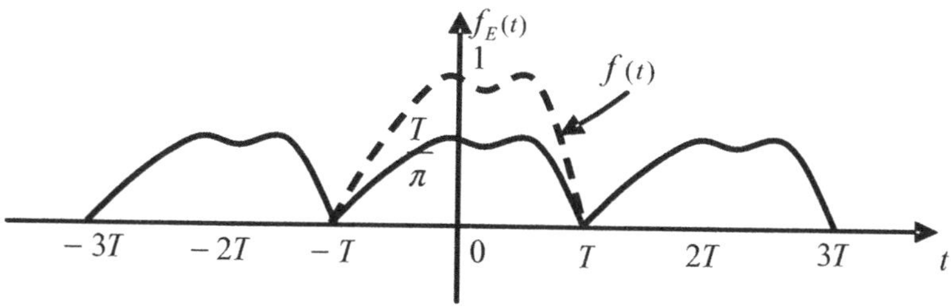

We can then write:

$$F(j\omega) = \frac{1}{2\pi}F_E(j\omega) * \frac{\pi}{T}2T\frac{\sin\omega T}{\omega T}$$

Thus:
$$\begin{cases} F(j\omega) = F_E(j\omega) * \dfrac{\sin\omega T}{\omega T} \\ \quad - \displaystyle\sum_{n=-\infty}^{n=+\infty} F(n\frac{\pi}{T})\delta(\omega \;\; n\frac{\pi}{T}) * \dfrac{\sin\omega T}{\omega T} \end{cases}$$

From this we establish that:

$$\boxed{F(j\omega) = \sum_{n=-\infty}^{n=+\infty} F(n\frac{\pi}{T})\frac{\sin(\omega T - n\pi)}{(\omega T - n\pi)}}$$

6.3.6. *Exercise 6.6*

a) The relation between input and output can be written using the time form:

$$s(t) = \int_{-\infty}^{t} \left(f_E(x) - f_E(x - T)\right)dx \quad \text{with } T = \frac{1}{2f_M} = \frac{\pi}{\omega_M}$$

The transfer function represents the Fourier transform of the impulse response of the circuit. This can be obtained by applying a Dirac pulse at the input.

$$f_E(x) = \delta(x) \Rightarrow h(t) = \int_{-\infty}^{t} \left(\delta(x) - \delta(x - T)\right)dx$$

$$\text{If } 0 < t < T \Rightarrow h(t) = \int_{-\infty}^{t} \delta(x)dx = 1$$

$$\text{If } t > T \Rightarrow h(t) = \int_{-\infty}^{t} \delta(x)dx - \int_{-\infty}^{u>0} \delta(u)du = 0$$

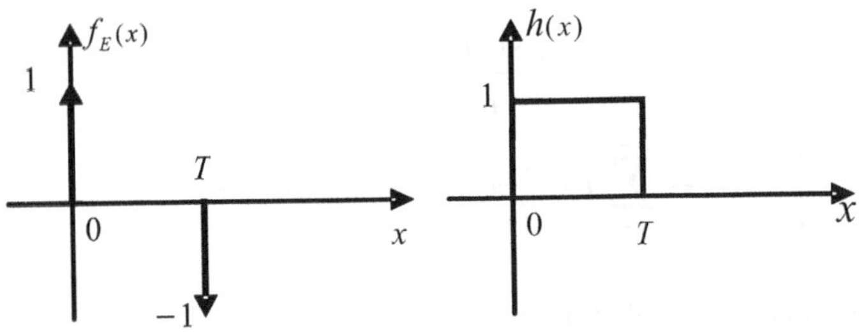

This allows us to establish the transfer function:

$$\boxed{H(j\omega) = Te^{-j\frac{\omega T}{2}} \frac{\sin\frac{\omega T}{2}}{\frac{\omega T}{2}}}$$

b) The sampled signal is written:

$$f_E(t) = \sum_{n=-\infty}^{n=+\infty} f(nT)\delta(t - nT)$$

The Fourier transform is worth:

$$F_E(j\omega) = \frac{1}{T}\sum_{n=-\infty}^{n=+\infty} F(\omega - n\omega_0)$$

Yet $S(j\omega) = H(j\omega)F_E(j\omega)$

Thus, if we replace this:

$$S(j\omega) = e^{-j\frac{\omega T}{2}} \frac{\sin\frac{\omega T}{2}}{\frac{\omega T}{2}} \sum_{n=-\infty}^{n=+\infty} F(\omega - n\omega_0)$$

We obtain the expression of a sampled and hold signal.

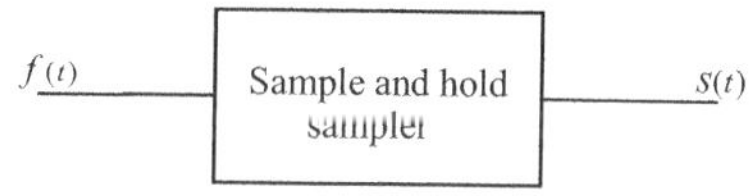

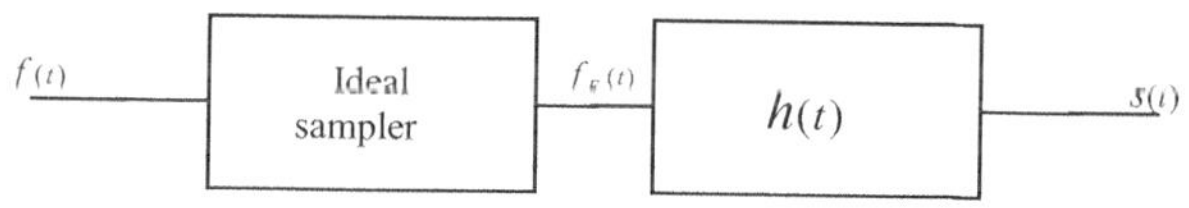

c) Let us plot the outlines of the various spectra:

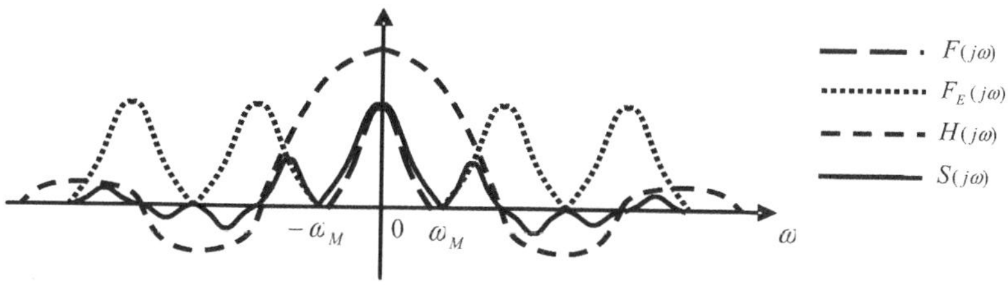

In frequency band $[-\omega_M,+\omega_M]$, transfer function $H(j\omega)$ has little impact on spectrum $F_E(j\omega)$. It reduces it heavily outside this band forces the amplitudes at points $n\omega_M$ to zero.

The result is that $s(t)$ is slightly different to $f(t)$.

d) We now trace the outlines of the different spectra with the circuit replaced by a low-pass filter of first order:

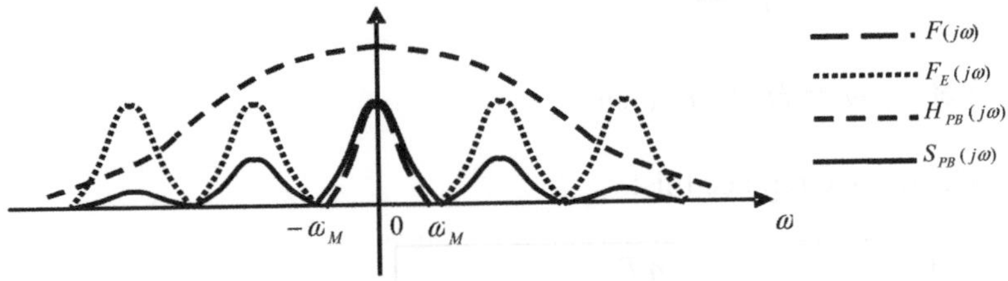

The impact of the low-pass filter is lower in the signal band but reduces the spectral components much less outside the band.

6.3.7. *Exercise 6.7*

a) Consider the following example:

$$\omega_M = 2.5\Delta\omega\,,\ \Delta\omega = 2rds^{-1} \text{ and } \omega_{\text{sample}} = 1.1\,\omega_M$$

After calculation: $\begin{cases} \omega_M = 5rds^{-1} \\ \omega_m = 3rds^{-1} \\ \omega_E = 5.5rds^{-1} \end{cases}$

We obtain the sampled spectrum by translating the initial spectrum by an integer multiple of the sampling frequency.

The outline of the sampled spectrum is the following for translations $\pm\omega_E$ and $\pm 2\omega_E$:

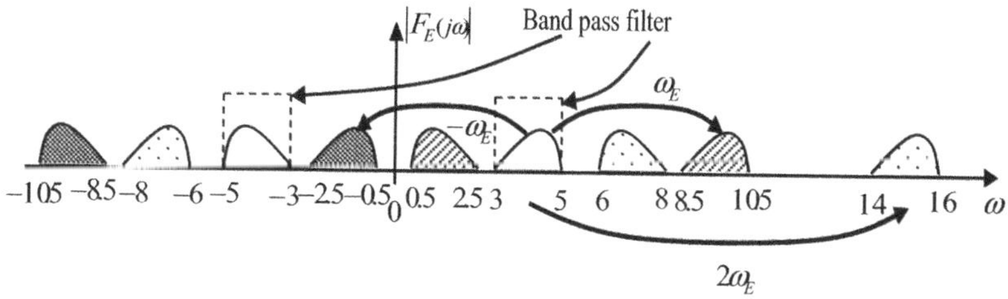

We can clearly see that this sampling frequency causes no spectral overlap and, through band-pass filtering, will allow us to reproduce the original signal. This is true even though the sampling frequency is lower than Shannon's frequency $2\omega_M$.

However, we will note that this result is not always guaranteed. If the sampling frequency becomes $\omega_E = 1.3\omega_M$ meaning $\omega_E = 6.5rds^{-1}$ we notice that the inferior band of the initial shifted spectrum of ω_E will partially overlap with the top band of the spectrum of the original signal. We can then no longer recreate the signal.

b) We will now analyze a generic situation:

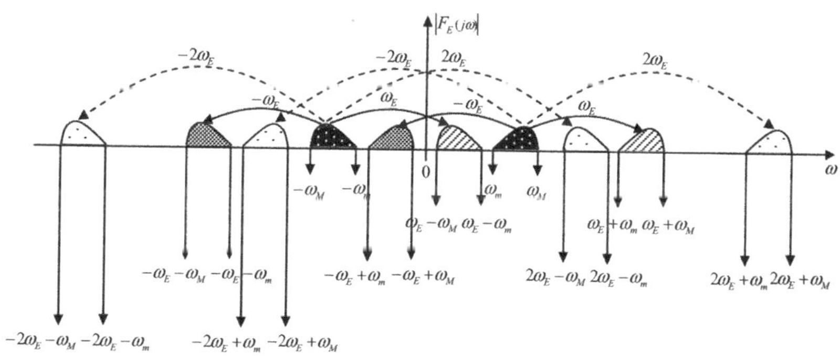

As there is no spectral overlap there will be a certain number of conditions imposed upon the frequencies of the translated spectrum. Only the two first translations need to be analyzed.

We obtain the following results:

$$\begin{cases} \omega_E - \omega_m < \omega_m \\ 2\omega_E - \omega_M > \omega_M \\ \omega_E - \omega_M > -\omega_m \end{cases}$$

After developing this:

$$\begin{cases} \omega_{E\,m} < 2\omega_m \\ \omega_E > \omega_M \\ \omega_E > \omega_M - \omega_m = \Delta\omega \end{cases}$$

The second inequality leads us to the third, giving us the following:

$$\boxed{\omega_M < \omega_E < 2\omega_m}$$

This result is only possible if $\dfrac{\omega_M}{\omega_m} < 2$.

The bandwidth must therefore be lower than the octave.

6.3.8. *Exercise 6.8*

Let us plot the spectrum of a periodic function containing a limited number of components.

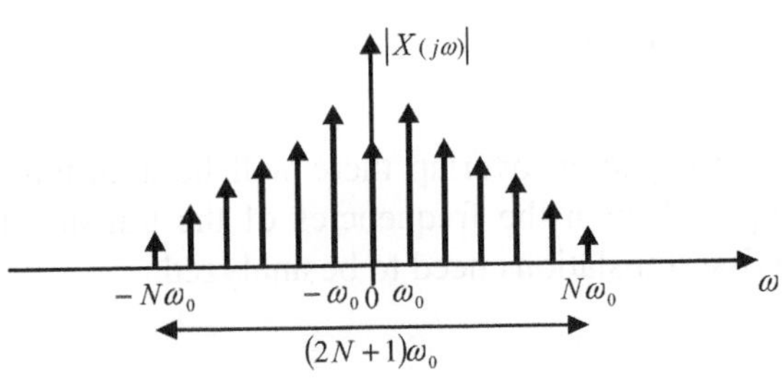

Let us now outline the spectrum of the sampled signal to a frequency equal to that of Shannon's.

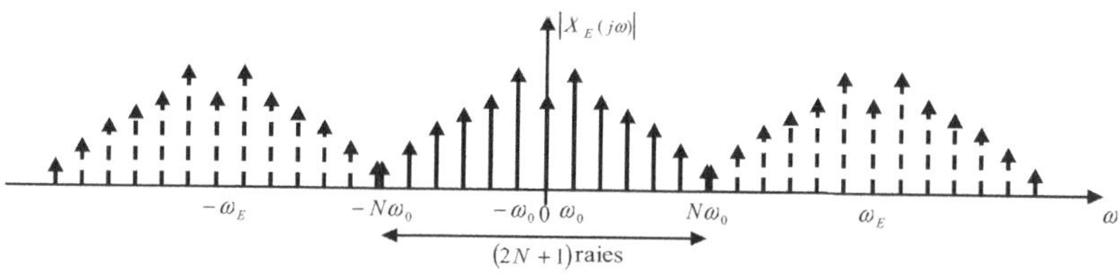

The discreet nature of the spectrum highlights the fact that simply knowing the value of the samples allows us to recreate function $x(t)$ entirely.

Mathematically speaking, this is the same as multiplying the sampled spectrum by a finite series of Dirac pulses:

$$X(j\omega) = X_E(j\omega)\sum_{k=-N}^{k=+N}\delta(\omega - k\frac{2\pi}{T})$$

We deduce from this the inverse Fourier transform:

$$x(t) = x_E(t) * \frac{1}{2\pi}\sum_{k=-N}^{k=+N} e^{jk\frac{2\pi}{T}t}$$

The expression of the sampled signal is:

$$x_E(t) = \sum_{k=-N}^{k=+N} x(kT_E)\delta(t - kT_E)$$

If the sampling frequency is such that:

$$\omega_E = (2N+1)\omega_0$$

Then:

$$T_E = \frac{T}{2N+1}$$

We notice that the term $\sum_{k=-N}^{k=+N} e^{jk\frac{2\pi}{T}t}$ is the sum of a geometric progression. We then obtain:

$$\sum_{k=-N}^{k=+N} e^{jk\frac{2\pi}{T}t} = \frac{\sin(2N+1)\frac{\pi t}{T}}{\sin\frac{\pi t}{T}}$$

We deduce the following:

$$x(t) = \sum_{k=-N}^{k=+N} x(kT_E)\delta(t-kT_E) * \frac{1}{2\pi}\frac{\sin(2N+1)\frac{\pi t}{T}}{\sin\frac{\pi t}{T}}$$

And obtain the desired result:

$$\boxed{x(t) = \frac{1}{2\pi}\sum_{k=-N}^{k=+N} x(k\frac{T}{2N+1})\frac{\sin(2N+1)\frac{\pi}{T}\left(t-k\frac{T}{2N+1}\right)}{\sin\frac{\pi}{T}\left(t-k\frac{T}{2N+1}\right)}}$$

Bibliography

[BEN 95] BENICHOU P., *Séries de Fourier. Transformation de Laplace*, Ellipses, 1995.

[BLA 01] BLANCHET G., GARBIT M., *Signaux et images sous Matlatab*, Hermes, 2001.

[COB 88] COBLENTZ J-F., *Introduction à l'analyse de Fourier*, Eyrolles, 1988.

[DEM 95] DEMENGEL G., *Distributions et applications*, Ellipses, 1996.

[LAT 66] LATHI B.P., *Signals, Systems and Communications*, John Wiley and Sons, 1966.

[LES 12] LESFARI A., *Distributions, analyse de Fourier et transformation de Laplace*, Ellipses, 2012.

[MOK 00] MOKHTARI M., *Matlab et Simulink pour étudiants et ingénieurs*, Springer, 2000.

[PET 98] PETIT J., *L'outil mathématique pour la physique*, 5th ed., Dunod, 1998.

[PEY 12] PEYRIERE J., *Convolution, séries et intégrales de Fourier*, Eyrolles, 2012.

[REI 02] REINHARD H., *Eléments de mathématiques du signal*, 3rd ed., Dunod, 2002.

[ROD 93] RODDIER F., *Distributions et transformation de Fourier*, Ediscience, 1993.

Index

H, I, L

M, O, P

R, S

T, U, V

Other titles from

iSTE

in

Digital Signal and Image Processing

2016

CHARBIT Maurice
Digital Signal Processing with Python Programming

FEMMAM Smain
Fundamentals of Signals and Control Systems

FEMMAM Smain
Signals and Control Systems: Application for Home Health Monitoring

2015

BLANCHET Gérard, CHARBIT Maurice
Digital Signal and Image Processing using MATLAB®
Volume 2 – Advances and Applications: The Deterministic Case – 2nd edition
Volume 3 – Advances and Applications: The Stochastic Case – 2nd edition

CLARYSSE Patrick, FRIBOULET Denis
Multi-modality Cardiac Imaging

GIOVANNELLI Jean-François, IDIER Jérôme
Regularization and Bayesian Methods for Inverse Problems in Signal and Image Processing

MAÎTRE Henri
From Photon to Pixel: The Digital Camera Handbook

2014

AUGER François
Signal Processing with Free Software: Practical Experiments

BLANCHET Gérard, CHARBIT Maurice
Digital Signal and Image Processing using MATLAB®
Volume 1 – Fundamentals – 2nd edition

DUBUISSON Séverine
Tracking with Particle Filter for High-dimensional Observation and State Spaces

ELL Todd A., LE BIHAN Nicolas, SANGWINE Stephen J.
Quaternion Fourier Transforms for Signal and Image Processing

FANET Hervé
Medical Imaging Based on Magnetic Fields and Ultrasounds

MOUKADEM Ali, OULD Abdeslam Djaffar, DIETERLEN Alain
Time-Frequency Domain for Segmentation and Classification of Non-stationary Signals: The Stockwell Transform Applied on Bio-signals and Electric Signals

NDAGIJIMANA Fabien
Signal Integrity: From High Speed to Radiofrequency Applications

PINOLI Jean-Charles
Mathematical Foundations of Image Processing and Analysis
Volumes 1 and 2

TUPIN Florence, INGLADA Jordi, NICOLAS Jean-Marie
Remote Sensing Imagery

VLADEANU Calin, EL ASSAD Safwan
Nonlinear Digital Encoders for Data Communications

2013

GOVAERT Gérard, NADIF Mohamed
Co-Clustering

DAROLLES Serge, DUVAUT Patrick, JAY Emmanuelle
Multi-factor Models and Signal Processing Techniques: Application to Quantitative Finance

LUCAS Laurent, LOSCOS Céline, REMION Yannick
3D Video: From Capture to Diffusion

MOREAU Eric, ADALI Tulay
Blind Identification and Separation of Complex-valued Signals

PERRIN Vincent
MRI Techniques

WAGNER Kevin, DOROSLOVACKI Milos
Proportionate-type Normalized Least Mean Square Algorithms

FERNANDEZ Christine, MACAIRE Ludovic, ROBERT-INACIO Frédérique
Digital Color Imaging

FERNANDEZ Christine, MACAIRE Ludovic, ROBERT-INACIO Frédérique
Digital Color: Acquisition, Perception, Coding and Rendering

NAIT-ALI Amine, FOURNIER Régis
Signal and Image Processing for Biometrics

OUAHABI Abdeljalil
Signal and Image Multiresolution Analysis

2011

CASTANIÉ Francis
Digital Spectral Analysis: Parametric, Non-parametric and Advanced Methods

DESCOMBES Xavier
Stochastic Geometry for Image Analysis

FANET Hervé
Photon-based Medical Imagery

MOREAU Nicolas
Tools for Signal Compression

2010

NAJMAN Laurent, TALBOT Hugues
Mathematical Morphology

2009

BERTEIN Jean-Claude, CESCHI Roger
Discrete Stochastic Processes and Optimal Filtering / 2nd edition

CHANUSSOT Jocelyn *et al.*
Multivariate Image Processing

DHOME Michel
Visual Perception through Video Imagery

GOVAERT Gérard
Data Analysis

GRANGEAT Pierre
Tomography

MOHAMAD-DJAFARI Ali
Inverse Problems in Vision and 3D Tomography

SIARRY Patrick
Optimization in Signal and Image Processing

2008

ABRY Patrice *et al.*
Scaling, Fractals and Wavelets

GARELLO René
Two-dimensional Signal Analysis

HLAWATSCH Franz *et al.*
Time-Frequency Analysis

IDIER Jérôme
Bayesian Approach to Inverse Problems

MAITRE Henri
Processing of Synthetic Aperture Radar (SAR) Images

MAITRE Henri
Image Processing

NAIT-ALI Amine, CAVARO-MENARD Christine
Compression of Biomedical Images and Signals

NAJIM Mohamed
Modeling, Estimation and Optimal Filtration in Signal Processing

QUINQUIS André
Digital Signal Processing Using Matlab

2007

BLOCH Isabelle
Information Fusion in Signal and Image Processing

GLAVIEUX Alain
Channel Coding in Communication Networks

OPPENHEIM Georges *et al.*
Wavelets and their Applications

2006

CASTANIÉ Francis
Spectral Analysis

NAJIM Mohamed
Digital Filters Design for Signal and Image Processing

Printed and bound by CPI Group (UK) Ltd, Croydon, CR0 4YY
16/09/2026
14963072-0003